SPRINGTIME

at

Cannon Hall
Farm

SPRINGTIME
at
Cannon Hall Farm

The Nicholson Family

**EBURY
SPOTLIGHT**

1 3 5 7 9 10 8 6 4 2

Ebury Spotlight, an imprint of Ebury Publishing
20 Vauxhall Bridge Road
London SW1V 2SA

Ebury Spotlight is part of the Penguin Random House group of companies
whose addresses can be found at global.penguinrandomhouse.com

Penguin
Random House
UK

First published by Ebury Spotlight in 2022

www.penguin.co.uk

A CIP catalogue record for this book is available from the British Library

ISBN 9781529149012

Printed and bound in Great Britain by Clays Ltd, Elcograf S.p.A.

The authorised representative in the EEA is Penguin Random House Ireland,
Morrison Chambers, 32 Nassau Street, Dublin D02 YH68

Penguin Random House is committed to a
sustainable future for our business, our readers
and our planet. This book is made from Forest
Stewardship Council® certified paper.

To all of our loyal supporters

For me, spring means lighter, warmer days – and lots of bloody hard work.

—Robert Nicholson

Contents

Foreword

On a freezing cold Friday morning in early January 2021, a chorus of tawny owls provided the atmospheric soundtrack as Robert Nicholson slid back the sturdy sliding door to Farrowing House Number One. The second national lockdown was in full force. All around the farm there was a crunchy dusting of snow on the ground and Robert's inquisitive young sheepdog Pip was enjoying a good sniff around. As usual, farmer Robert was in an upbeat mood.

Filming had already begun for the new series of the popular Channel 5 show *Springtime on the Farm* and Robert was getting ready to present one of his regular live broadcasts to Cannon Hall Farm supporters via social media. He turned his phone to selfie mode, pressed the button to go live, and off he went.

"Good morning. I'm not going to speak too loudly as the pigs have got out of the habit of people being around, so I'm just hanging back a little so that Pip and I don't make them jump. It's a beautiful morning here and we're on the right side of winter now. A couple of weeks beyond the shortest day of the year and we're just beginning to see the difference. We're on a journey to better times as the vaccine is rolled out and I'm feeling positive for the future."

No matter what's going on in the outside world, as any true Cannon Hall Farm fan will tell you, this little oasis in the heart of South Yorkshire has become an anchor in troubled times – and not just for the members of the Nicholson family who live there. With an ever-growing network of supporters who regularly tune in to Robert's broadcasts and millions of TV fans all around the world, catching up with life on Cannon Hall Farm is a real tonic for everyone.

And in spring, life really steps up a gear . . .

Introduction

A Farming Dynasty

On 9 April 2018, the television programme *Springtime on the Farm* made its debut on Channel 5, celebrating the joys of the British countryside during springtime. It introduced us to the Nicholson family, the farmers who live and work at Cannon Hall Farm, near Barnsley in South Yorkshire.

Roger Nicholson heads up the family and comes from a farming dynasty that can be traced back to the 1600s. He has lived at Cannon Hall Farm ever since his father bought the 126-acre property at auction in 1958. Prior to their big move, the Nicholsons lived in Bank End Farm in the nearby village of Worsbrough Dale, where several generations of the family worked the land and made their mark in the farming world. The farm was passed down from brother to brother and from father to son, each of them maintaining the land, growing healthy crops and caring for their animals.

Roger's charismatic father, Charlie, was himself a highly respected stockman and made headlines with his prize-winning white bull, whose pedigree name was Fockerby Ringleader. As well as being a fine Shorthorn specimen, the bull was partial to a drink of beer and would happily swallow down a pint after a country show victory, much to the amusement of spectators. The story actually made the local newspaper; even back then Charlie Nicholson could recognise the marketing potential of an interesting animal on a local farm.

Although they had lived at Bank End Farm since 1798, the family rented most of their land. And what started out as a fairly sizeable farm of 250 acres began to shrink as the land was sold off bit by bit because of compulsory purchase orders. Acres of beautiful farmland full of crops were sold off and developers turned it into a housing estate and a local school. By the end of their time there, only 30 acres of the original plot remained, which, for the Nicholsons, wasn't viable to sustain a family business. A fresh start was needed.

Roger was still at school when the family moved to their new home at Cannon Hall Farm, and, although he intended to take over the family business one day, fate intervened when Charlie died suddenly of a heart attack in 1959. Soon after, Roger, then aged just 16, had to leave school and quickly learn how to make a living as a farmer so that he could support himself and his mother, Rene.

Three years later, Roger met his future wife, Cynthia Dickin, at a Young Farmers' Club dance in Halifax. The story goes that while feisty Cynthia greatly impressed Roger by winning a competition at the dance, she wasn't at all enamoured by his advances and it took a bit of convincing for her to agree to meet him again. But something must have won her over and, three years of courtship later, Roger and Cynthia married in 1965 and moved to the picture-perfect Tower Cottage at Cannon Hall Farm. Their eldest son, Richard, was born in 1966, followed by Robert in 1968 and youngest son, David, two years later.

It was an idyllic environment for the Nicholson boys to grow up in, surrounded by acres of lush farmland to play in, woods to build their secret dens and streams to catch fish. In the early years, Roger kept a variety of livestock, including dairy cows, pigs, sheep and some hens, but, although he worked as hard as he was able, he struggled to make a living from farming. On the days when he'd take his animals off to be sold at market, he never seemed to get the rightful return for all his hard graft. The cost of rearing animals to enable him to produce really top-quality meat was never reflected when deals were struck and the hammer came down. And later, as bigger farms became more industrialised and smaller homesteads were effectively squeezed out of the marketplace, the Nicholson family knew that they had to diversify their farm output so that they could keep a roof over their heads.

Living within your means is one thing, but when it's year after year of struggle, and others who started out on the same level playing field seem to be getting more and more success-ful, it's time for a rethink. Not that Roger was ever one to rest on his laurels; he'd focus on expanding a breed, then make strides in a different direction, hoping each time that he would settle on the most successful direction for the farm. For example, when his dairy herd wasn't profitable enough, he started rearing more pigs to see if that was where he'd find bet-ter returns. Each time, his efforts would pay off initially, but it was never quite enough for the farm to become really success-ful, or for life to get any easier.

Cannon Hall Farm was once part of the Cannon Hall Estate, owned by the Spencer-Stanhope family, who made their for-tune in the iron industry. In the 1950s, the estate was broken up and sold, and the original Georgian country house was later turned into a museum. The stunning sandstone house stands proudly just outside the village of Cawthorne, about five miles west of Barnsley, and is surrounded by 70 acres of beautiful parkland and gardens. The museum, park and gardens are owned by Barnsley Council, and have always been popular with visitors. Being just across the courtyard from the Nicholsons' farm, the boys would often explore the surroundings and make mischief there – much to the annoyance of head gardener, Mr Hales.

Back in the early 1980s, Cannon Hall Museum didn't have a café for visitors, so, figuring that they could fill a hungry gap in that particular market, the Nicholsons came up with the idea of converting one of their farm outbuildings into a traditional country tea room. It just so happened that Cynthia was a dab hand at baking, and she and her friend Rosemary could head up the new venture. There were barriers to overcome – not least from the museum itself – but in April 1981, Home Farm Tea Rooms at Cannon Hall Farm opened to the public for the first time. Cynthia's mum, Olive, was on washing-up duty, and Cynthia and Rosemary worked long and hard to produce dozens of scones, meringues, cakes and savouries. With a full cream tea costing the bargain price of £1.45, it didn't take long until word got around the local area about how good it was, and the customers poured in. Bigger premises were soon needed. The family all mucked in to help, with the now-teenage Nicholson sons waiting on tables and working the till.

Meanwhile, several farms up and down the country had opened up to the public, and the Nicholsons wondered if Cannon Hall Farm could thrive as both a working farm and a visitor attraction. While the farmers could go about their day, working their crops and rearing livestock, the public could watch them at work, see milking demonstrations, hear about the various breeds and feed and pet some of the smaller animals – the ones that weren't likely to give them a nasty bite . . . As a nation of pet lovers, there seemed to be a real appetite for the chance to handle a baby piglet, feel just how soft an Angora rabbit is and watch a newborn lamb being hand-fed. It took the zoo idea to the next level. The Nicholsons visited several similar farms up and down the country and started making plans. By investing in a few new animals and borrowing ones from here and there, they would, they believed, be able to create the perfect open-farm experience. It would offer a mix of education and entertainment, so it would also be a valuable resource for schools.

Like every major project, it wasn't a venture that could happen overnight, and it required a hefty investment to make it happen. By gradually selling off some of the farm property, converting outhouses into animal barns and eventually, at virtually the last minute, managing to secure a vital expansion loan from the bank, plans could go ahead to transform the humble farm into a dynamic visitor attraction.

It looked a bit rough and ready, but on Friday 24 March 1989, Cannon Hall Farm was officially opened to the public, just in time for the Easter holidays. Although the family was happy with their initial takings – after all, it was money coming in rather than going out – there were soon plans afoot to make it bigger and better. Like the Home Farm Tea Rooms, such was the success of the open farm that they quickly needed to expand to better accommodate visitors. New barns and paddocks were constructed to house more animals; Roger invested in new breeds; and better facilities for the public were created, including a gift shop.

And then there was the car park. Who would have imagined that such a basic thing would involve so much bureaucracy and red tape? Or that putting in a new toilet block would cost a king's ransom? To enable the Nicholsons to bring their grand plan to fruition, all of these hurdles had to be crossed, with more head-scratching headaches and endless lists of figures. Thankfully, there were more glamorous facilities that opened too, such as the Nicholsons' farm shop delicatessen, which opened in 2005, plus an exciting adventure playground and

The Hungry Llama – a new family restaurant, complete with a children's soft play area. Two years later – to celebrate Roger's father Charlie and his prize-winning Shorthorn bull – the impressive White Bull restaurant was opened. What was once just a local country farm was now one of Yorkshire's most successful tourist attractions.

The open farm was a family affair from the word go, with Richard, Robert and David all mucking in to help with the building work while the transformation was taking place. This meant taking a sledgehammer to some of the barns and building from the foundations up – complete with some rather uneven but "it'll do for now" flooring. When the brothers left school, Robert and David honed their skills as farmers at college, while Richard studied photography and worked on the marketing side of the business, helping to further spread the word about Cannon Hall Farm. The whole family was determined to evolve the farm and do everything they could to make sure that it flourished and became the success story that it was always destined to be.

Visitors could feel right at the heart of farming. The Nicholsons quickly grew accustomed to having members of the public watch in awe from the sidelines as the farmers helped to deliver hundreds of new baby animals during springtime. Later, they built farrowing barns with viewing balconies so that the visitors could also watch lines of newborn piglets hungrily suckling from their mums. Next came the excellent Roundhouse with its 360-degree mezzanine viewing platform.

It wasn't long before regional news teams heard about the public's love of seeing lambing at Cannon Hall Farm and came along to film the Nicholson farmers in action. In May 2011, the farm stayed open to the public all night so that visitors could watch the lambing close-up. In footage from *BBC Look North*, middle son Robert explained the thinking behind allowing cameras to capture them at work: "It's what made us want to do it in the first place; we love to tell the story of where our food is produced and we think we have the ultimate in provenance for the things we produce."

As smartphone technology stepped up a gear, people also started to capture farm events live on their mobile phones, and Cannon Hall Farm's fame began to spread far and wide. In 2015, Robert began broadcasting each Sunday via Facebook and the farm's social media following grew like wildfire. He'd report on new arrivals, update viewers on how the various animals were enjoying life at the farm and fill them in on any new funny or sad stories that had occurred. Soon a whole community of supporters came together. They're still loyal fans to this day and are an important part of Cannon Hall Farm's continuing success. "Originally we only did one-minute broadcasts because we thought that people might get a bit bored," says Richard. "But they seem to have gone on a bit longer since then . . ."

In 2017, the award-winning TV production company Daisybeck got in touch with the family about a new Channel 5 show that was due to air in the spring of the following year. *Springtime on the Farm* was to be a celebration of farming throughout the early months of the year, and the production team was looking for a farm where the live show would be based. There would be reports from all around the country, plus footage of spring events happening at the host farm. Two presenters would anchor the show and celebrity guests would roll up their sleeves and muck in with the farming.

Daisybeck had already had great success with their awe-inspiring programme *The Yorkshire Vet*, following the work of the vets at Skeldale Veterinary Centre in Thirsk (at which Alf Wight, better known under his pseudonym of James Herriot, had worked) so they knew there was a massive appetite for factual animal stories set against the backdrop of the stunning Yorkshire countryside. And after seeing the fantastic combination of Cannon Hall Farm's beautiful location and recognising the integrity of the Nicholson family, the production company knew they were on to a winner.

Television audiences would be able to witness the first moments of a newborn lamb and see the hard work, sheer grit and determination needed in order for a successful springtime to unfold on a farm. The programme would be an insight into how farm animals are reared from the moment they take their first breath and would reveal how farms up and down the country tackle the challenges that face them. In the opening show on 9 April 2018, Robert was filmed in action with his younger brother David demonstrating some of the techniques that farmers use to assist with difficult lambings. As the farm is also home to goats, cows and a menagerie of other animals, television viewers could also witness all manner of other springtime animal births.

The format proved so popular that *Springtime on the Farm* was back in 2019 and 2020. And even when the Covid-19 pandemic was in full force, the team found a way to broadcast the sights and sounds of rural springtime in the magical Channel 5 TV show. What's more, it seemed that television audiences wanted further updates from the Nicholson family and more action from the farming world, not just in springtime.

"The original vision for *Springtime on the Farm* was to focus on the human side of farming, and the farmers who dedicate their lives into putting food on the table at the busiest time of the farming year," says Paul Stead, the Managing Director of Daisybeck. "The third year of *Springtime on the Farm* was filmed during Covid-19, and Channel 5 loved it so much they asked the Nicholsons if they would like to carry on filming events at the farm and make a new series called *This Week on the*

Farm. Several of the scheduled Channel 5 programmes had been cancelled because of the pandemic, but, because we could successfully film the show remotely, it was a perfect fit at a time when people were hungry for a distraction from the pandemic. Each series follows the thrust of the seasons and we've covered them all in *This Week on the Farm*."

The television programmes based at Cannon Hall Farm mean even more people have shared in the joys of life there, which is now home to both traditional farm animals such as sheep, pigs and cattle, as well as farm favourites such as Shetland ponies like John Bon Pony, Pony Hadley and Ozzy Horsebourne, more than 60 pygmy goats, including Millie and Primrose, and a whole colourful cast of creatures great and small. Cannon Hall Farm was granted a zoo licence when their Reptile House was opened in 2017, and the farm is now also home to everything from thousands of tiny leafcutter ants to several mighty Shire horses, llamas, alpacas, meerkats, ferrets and even reindeer. Needless to say, it's certainly not the average Yorkshire farm.

As Cannon Hall Farm is both a working farm and a visitor attraction, the majority of pigs, sheep and cattle that are born on the farm are reared for meat. There are exceptions in the rare breeds – the Swiss Valais Blacknose sheep, for example, are sold on as breeding animals, as are any particularly good examples of other breeds. With a team of animal experts that they can call on, and by regularly assessing how each breed on the farm is handled and cared for, the Nicholson family continue to build on their ethos; as Robert says, "Whether you are a human or an animal, all you can hope for is that you have a good life. We are lucky enough to have breeding stock on our farm and we have some animals that are glorified pets, but with our farm animals that we rear for the farm shop, we like to think that we are helping them live their best life."

Each new year brings new arrivals; hundreds of lambs are born as well as several litters of pigs, goat kids and calves, plus plenty of other furry, feathered, two-, four-, six- and eight-legged friends. Between them, the animals and farmers have captured the nation's hearts and put Cannon Hall Farm firmly on the map.

THE
NICHOLSON
FAMILY

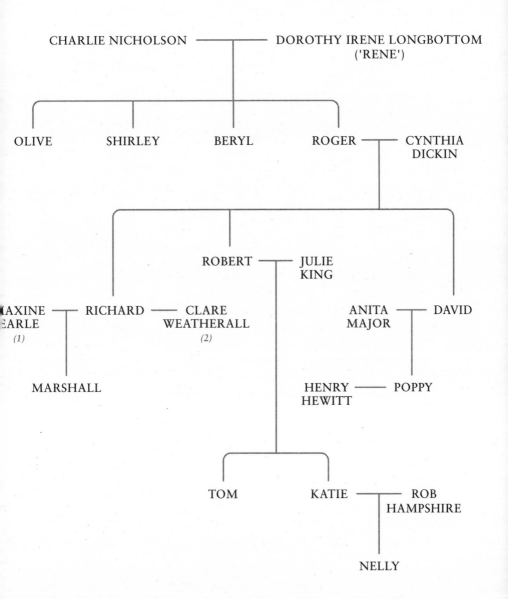

CHARLIE NICHOLSON ———— DOROTHY IRENE LONGBOTTOM
('RENE')

OLIVE   SHIRLEY   BERYL   ROGER ——— CYNTHIA DICKIN

ROBERT ——— JULIE KING

MAXINE ——— RICHARD ——— CLARE   ANITA ——— DAVID
EARLE   WEATHERALL   MAJOR
(1)   (2)

MARSHALL   HENRY ——— POPPY
            HEWITT

TOM   KATIE ——— ROB HAMPSHIRE

NELLY

PART ONE
January

Tasks for January
Clean up turkey barns
Repair any farm machinery
Fix drystone walls
Check fences, hedgerows and boundaries
Other general maintenance – frozen pipes? Leaking roofs?
Harvest kale and turnips
Scanning, vaccinating and supplementary feeding of sheep due
  to lamb in February
Organise lambing supplies – feeding equipment, first aid, heat
  lamps, ropes, iodine, etc.
Scheduled calving, farrowing and other animal arrivals

1

New Year's Day can't help but hold the promise of a fresh start. Even if there's a legacy of a bad year that you want to leave behind, the fact it's the first day of January is a good reason to take a deep breath and step forward into a new page of your life; to wipe the slate clean, maybe make a New Year's resolution or two and indulge in a dose of positivity.

For the Nicholson family at Cannon Hall Farm, the optimism of a new year is always something to celebrate. Some may be a little weary from excessive New Year's Eve celebrations, but work goes on as there are jobs to be done 365 days a year. Just as we all need to make sure our families are nourished, happy, healthy and comfortable every day, the same goes for the animals on Cannon Hall Farm. They are all part of the family, after all.

Over the years, the Nicholsons have faced all manner of adversity and battled plenty of personal and professional challenges, but they have always come out stronger and more resilient. There have been plenty of reasons to celebrate and the family has continued to grow, with new generations to add to the story; all those milestones such as first words and first steps, the announcements of new jobs, weddings and children – as well as occasions when luck seems to deal one bad hand after another.

Through all this, Cannon Hall Farm has diversified from a traditional family concern to a major business that now provides a living for hundreds of staff. Along with a workforce of farmers who care for the animals, there are dynamic teams of retail staff, chefs, restaurant workers, marketing experts, strategists and all the key workers that are needed for the business to run like a well-oiled machine. It's constantly innovating and moving forward. In the 30 years since Cannon Hall Farm has been open to visitors, it has seen visits from royalty, won enough awards to fill several mantelpieces and is celebrated each and every day on social media with fans all around the world. Not bad for a little farm that started out with a few cows, sheep and pigs. Each new year is another chapter in its exciting story.

Although 1 January is just another day in farming terms, with all the same day-to-day tasks to be carried out, it's one step closer to spring, the busiest time in the farming year. From Land's End to John o' Groats, no two farms will operate in the same way, and while there may be similarities in the way that things are done, different climates dictate variations in the timing of tasks and the choice of breeds to nurture and crops to grow. It's all about getting the most out of your land and working with nature, the seasons and – possibly most important of all – the weather conditions.

"For me, springtime is when the farm awakens after winter," says Roger Nicholson, who has lived and worked at Cannon Hall Farm for over 60 years. "There's a bit of a bite in the grass, the trees come into bud and the spring flowers start to push through the soil. Every year, spring seems to arrive on a different date and, on New Year's Day, while it's still a little while off, it's a great thing to work towards. It's certainly a reason for us to all feel positive."

Roger, Cynthia and their sons, Richard, Robert and David, are the host farming family featured in Channel 5's *On the*

Farm series. Since 2018, they have been delighting TV viewers with tales of life from their South Yorkshire farm. Viewers
get to see the day-to-day work – and sometimes grind –
involved in running a working farm, and also regularly catch
up with some of their animals, who have become stars in
their own right. Who could forget Zander, the irrepressible
alpaca, or Stanley, the psychic goat? There are the llamas
Elvis and Priscilla, Gary the randy donkey, Fern, Ted and
the Highland fold posse, reindeer Prince and Jeffrey, and
many, many more.

Every year on the farm, there are highs and lows, as fans
share not only the magical moments, but tragic times too. It
makes your heart sing to see a new foal galloping in the paddock for the first time in its life, or when a new batch of pygmy
goat kids discover the joys of their very own adventure playground. Then there are those terribly sad times when a beautiful
baby donkey isn't strong enough to survive or a traumatised
cow has a stillborn calf.

Along with all the animal antics, Robert and David are
always eager to try their hand at traditional farming skills,
country crafts and physical challenges. Their competitive spirit
makes for great viewing as they battle it out to be the best at
everything, from gin-making and scone-baking to axe throwing
and welly wanging. Life on a farm doesn't get more colourful
than this.

Thanks to social media, fans can also keep up to date with all
the other goings-on at the farm, such as special events and new
facilities, announcements and activity timetables. As the farm
is now a tourist attraction as well as a working farm, visitors
know that they can come back again and again and there will be
new things to see and attractions to fill their day, from watching
real-life woolly jumpers going head-to-head in the sheep races
to learning about new breeds as the farm continually introduces

new animals to its line-up. Who would have thought that you could find a pair of porcupines living happily near Barnsley? Where other open farms may enlighten and entertain, Cannon Hall Farm really goes the extra mile.

With 126 acres of farmland at the main Cannon Hall Farm site and around 60 acres at the family's other farm in the nearby Gunthwaite Valley, there's ample space for their flocks of sheep and cattle to graze and for their collection of traditional farm animals and exotic breeds to thrive. It's a farm for all seasons, so animals can be sheltered from harsher weather conditions when necessary and live comfortably outside when the climate is kinder. The farm is situated at the foothills of the Pennines, so it's seldom battered by the worst rainfall or the most dramatic snowfalls that some areas of the north of England can be dealt. Nevertheless, over the years, bad weather conditions have been a real challenge to the Nicholsons – sometimes with heartbreaking consequences.

Understandably, the vagaries of British weather are a constant cause of concern for farmers. Livelihoods depend on the climate for optimum crop growth and for farm animals to flourish. And whether farmers have seemingly endless acres of land at their disposal for their flocks and herds to roam and graze, or a more contained area to manage, extremes of weather are a constant concern. A bountiful crop that has been nurtured since the day it was planted can be ruined in a matter of hours by freak weather. Extremes at either end of the barometer can spell disaster for farm animals as temperature fluctuations lead to cases of pneumonia, other respiratory infections and even death.

Although we joke about the weather being a national obsession, farming is so weather dependent that the subject is more serious than mere small talk. There's truth in old farming sayings such as "red sky at night, shepherd's delight; red sky in the morning, shepherd's warning". A rich red sky is a result of dust and soot particles being trapped by high pressure. While air molecules scatter the short blue wavelengths of sunlight, the trapped dust and soot scatter the longer red wavelengths. This is called Rayleigh scattering. A red sky at night indicates the high pressure and better weather is westwards, so on its way. But as morning light comes from the east, if there's a red sky it means the good weather has already passed and a red sky in the morning means low pressure – and bad weather is following.

"Make hay while the sun shines" is less scientific. It simply conveys the fact that hay needs to be nice and dry when it's harvested to avoid mould. Meanwhile, the saying "when the wind is out of the east, it's no good to man nor beast" refers to chill winds originating from Siberia that can freeze the country to the bone.

"A year of bad weather on a farm is horrendous," says Roger. "You can make no money and have nothing more to show for it than a pile of grief. Equally, in a good year, you can have a spring in your step and a smile on your face. Weather does play a big part, but you have to try not to be either too down about it or to get too optimistic – try to find a happy medium. Luckily, over the years I've got slightly better and don't tend to stress as much over the weather.

"We've had years when it never seems to stop raining for weeks, which makes it impossible to turn out the cattle into the fields," says Roger. "They'd hate it and it would wreck the fields so we just have to keep them inside. The cold isn't so much of a problem, but snow and ice bring all sorts of extra challenges. If your animals are out on the fields, you have to think of their safety, and even being able to get out to feed them can be a problem when the conditions on the roads are treacherous. Equally, when it suddenly gets very warm, animals can pick up bugs more easily. It's a balancing act, really, and you just have to take things day by day."

2

Born in March 1943, Roger has seen a lot of changes in the way that springtime unfolds year after year. Ask anyone over a certain age and they are sure to say that seasons aren't so defined any more – that there always used to be a clear distinction between winter, spring, summer and autumn.

For farmers like Roger, this means noticing how nature dictates necessary changes, and adapting one's ways of working with the livestock to fit those. The Nicholsons also adapt the farming calendar to work for them. For example, traditionally lambing used to begin on 1 April, but with the farm now open to the public, these days the year's first delivery of lambs at Cannon Hall Farm is scheduled to take place in mid-February to coincide with the local schools' half-term holidays. This is followed by a busier lambing season at Easter.

When Roger was growing up, his family lived on a farm in the village of Worsbrough Dale, around eight miles away from the village of Cawthorne, where Cannon Hall Farm is situated. Back then, the family reared cows rather than sheep, so when a new year began, the focus would be on keeping the herd healthy while they wintered indoors.

"Depending on the weather, the cows would be brought in

for the winter around October every year, so by January they would be very happy and warm and well used to their indoor conditions," says Roger. "They certainly didn't want to be outside when it was freezing cold and there would be nothing for them to graze on. If you opened the barn door and it was pelting with rain they wouldn't try and make a break for it – they'd just look at you as if to say, 'Sod you, I'm staying indoors!' They were perfectly happy and there was always a special smell in the cow house – it wasn't objectionable, to be honest."

The farm at Worsbrough Dale had around 250 acres where the family kept a number of farm animals: 15-20 dairy Shorthorn cows; a dozen sows for pigging; and around a dozen calves that would be reared on the farm and sold on for meat production. The children also kept a few sheep and hens as pets. It was a traditional mixed farm as the Nicholsons also grew cereal crops, kale, turnips and potatoes, and that's how it had been for generations. Interestingly, local historical documents record that traditionally the family were flax growers, but this was before Roger's time.

Roger's wife Cynthia also grew up on a farm, but on the other side of Yorkshire to Barnsley, in Halifax, a thousand feet above sea level where winters could be much harsher. "In some homes you could go from the kitchen straight into the cow shed, so the smell permeated through the whole house. But having your cows so close made the house warmer, so nobody really minded the smell. If you go further back, people used to live in homes above their cows because the rising heat from the cows kept them warm."

While little Roger was helping to take care of the herd in Worsbrough Dale, January would be a very busy time for Cynthia's family, who reared laying hens. "We kept Rhode Island Reds," she says, "and in order to have them hatching in spring, we'd put the cockerels in with the hens in early January. We had several nest boxes, and as they went in to lay their eggs, a door would flap down in each compartment to keep them there. All the hens were ringed, so you could identify which egg was laid by which hen. Every few hours my older brother Ted and I would go around collecting the eggs and carefully writing which number chicken they came from. Then we'd gently carry the eggs through to our incubator in a warm dry towel and wait for the magic to happen."

Cynthia recalls how, sometime later, the lights would be turned off in the incubator room and each egg would be gently held over a frame of lights – a process known as candling. If there was a dark patch inside the egg, it meant there was a chick developing in them. "Any that didn't have that dark shadow meant no chick, so they'd be picked out – and probably used to make a sponge cake. We never wasted anything – even though the fact that they had been kept in an incubator meant they wouldn't have been that fresh!"

As the chicks hatched, the family would keep some for the following year's batch. The majority, however, would be sold as day-old pullets (females) for more egg production. Chickens bred solely for meat production are a different variety and tend to grow more quickly. Egg-laying hens aren't as tender as hens raised for meat as they're kept for longer, plus they'd be tougher to eat as their muscles have had to work harder.

"As the chicks started to hatch, they'd be a lovely rich yellowy brown colour," says Cynthia. "When they grew bigger and started laying eggs, I remember the sound of them chirping and singing away. It was deafening in there! They ate from a cafeteria system so they had continual access to food and water and you could press a button and all the dirt would be hoovered up. It was fairly fancy for the time. They were certainly very happy – an unhappy hen won't lay eggs for long."

Food-wise, Roger and Cynthia did rather well when they

were growing up. Even though there was rationing throughout the war – and after it until 1954 – there was plenty of game to get them through winter and the end of the shooting season on 31 January. "Meat was rationed, but wildlife on the farm was plentiful, so there would always be rabbit, or pheasant. There was never that much meat on it, but it would be tasty," Roger says. "Mum used to save up her sugar rations and make toffee for me and my sisters and we would share out our sweet rations and make a quarter of liquorice last a month."

When the young Roger was still too little to help out at his farm in Worsborough Dale, he would toddle around his father "helping" as much as he could with the day-to-day tasks. His three big sisters – Olive, born in 1924; Shirley, born in 1929; and Beryl, born in 1931 – were significantly older so they were never his childhood playmates: "They were past it by the time I came along!" he says. But once Roger reached school age, and with a farm at his disposal, he and his primary school friends – Freddie, Billy and Michael – created their own adventures in the Worsbrough Dale countryside.

From the age of eight, Roger was put in charge of milking two of the family's Shorthorn dairy cows, Polly and Molly, every afternoon. They were considered the quieter cows of the family's herd, meaning they didn't kick out too much when they were being milked, so they were good for young Roger to practise his milking skills on. Later, as Roger reached his teens, he'd be helping out with other farm tasks that required attention 365 days a year, such as feeding, watering and mucking out the farm livestock and caring for any animals that were kept as pets. "In the winter I'd be mucking out the cows every day when they were inside. There was a step down in the cow house where all the muck would drop and it would just be a matter of shovelling it up every morning. Not a great job, but a necessary one."

The oats, wheat and barley that were needed to feed the

animals were grown on the farm, so this meant ensuring feed supplies were always topped up. "Often in January we'd need to have a threshing day," says Roger. "Through the winter we'd use up our supplies, so we'd need to thresh more to replenish our stocks. I still remember the heavy thumping 'Ffo-ffo-ffo' sound of the tractor pulling the huge threshing engine up the road and drawing up to the side of the stack."

The tractor would be positioned at the front of the thresher and a belt would be connected to drive its powerful mechanism. "It was incredibly dangerous because there was this belt flapping away and people ducking underneath it as they fetched and carried," says Roger. "I'm sure there must have been accidents at some stage, but everyone just got on with their job. There was no health and safety legislation back then!"

It was a full-scale operation with several people working on it at one time. Although farms would have their own day-to-day machinery, a major piece of kit like a threshing machine would be owned by a contractor. And in addition to the workers that came along to operate the machine, there'd be all hands on deck to make the most of each visit. It was a significant expense for the farmer, but couldn't be done any other way.

Farmhands would work in sync, throwing the sheaves from one person to another ready to be fed into the thresher. The machine would then separate the crop into grain, straw and chaff. The chaff would be mixed with the horses' feed, the straw was used for bedding and the grain was a valuable cereal, so nothing was ever wasted.

Several men were needed to pick up the sacks of grain as they filled, collect the chaff and stack bales of straw as they came out of the back of the thresher. "The hardest, heaviest job would be fetching and carrying the bags of grain because there'd never have less than two hundredweight in them [around 100kg]," says Roger. "They'd heave it on to their backs, carry it across the yard, up some steps and into the grain store. And by the time they got back to the thresher, there'd be another bag ready to go. I remember my friend Will Roe used to be able to carry a bag weighing twenty-four stone on his back – now that's proper weight lifting."

It was tough, physical work and would take as long as it needed to take to make the most of each visit. The machine would arrive first thing in the morning and the team would work through until light began to fade. No time for lunch hours in those days.

"It was wonderful to see the huge machine in action, but what was really exciting for us as kids was uncovering the rats nesting under the logs the grain was stored on," says Roger. "If it was a school day and the kids heard the threshing machine was coming up the road, as soon as they could they'd race to the farm to watch. There'd be straining at the leash to get out of school and there would always be this surge of kids when the school bell rang. As the stores got lower, the rats were disturbed and started scurrying around and the noise of the machine and the squeals of the kids would be deafening! There'd be a frenzy of activity as the kids ran around chasing the rats with big sticks. Everybody loved a bit of ratting and we got a hundred and thirty-six rats in the barn one day!

"We always had lots of mice too, but they lived high up on top of the stack and would scarper pretty quickly if disturbed. They wouldn't have been able to survive further down as the rats would kill them."

As well as cereal crops, the family also grew vegetables. Turnips and potatoes were stored in a triangular-shaped mound covered in straw and soil known as a pie, to keep them fresh. The idea looked basic – just a big pile of turnips or spuds, as long as it needed to be, around four feet high and several feet wide, and protected from the elements by a "crust" of straw and soil. As you needed them for cooking or to sell, you'd just take out your spade and dig into the pie. It meant that you could harvest your root vegetables when they were ready and you could use them through winter.

Roger remembers his dad Charlie making up pies and riddling the potatoes to remove the dirt, then he'd grade them for size, putting the smaller and damaged ones in a broiler for the pigs and selling the bigger ones around the nearby housing estate. "I can still hear his voice," he says. "Twenty-eight pounds a bob, pod-a-doze!" (A bob was 5p.)

"One particular job in January that was absolutely loathed was cutting kale," Roger says. Long before kale became a trendy vegetable, it was just another winter feed for animals and it had to be harvested by hand. "It was a horrendous task because sometimes in the depth of winter you'd have snow four inches deep and you'd no choice but to dig down and cut the kale as it was needed to feed the animals. We'd no gloves and your hands would be numb and blue with the cold, but each head of kale had to be individually cut at its root. Then it would all have to be picked up by hand and loaded on to the trailer. When there are icy winds in your ears and a whole field to harvest it just feels like a never-ending job. It was back-breaking, wet and heavy work."

Turnip picking also had to be done in winter. Again, by hand. Again, a laborious, relentlessly boring job, often in freezing cold weather. "It was a hell of a wet, mucky job. We'd be in the field filling the trailer, then we'd haul the turnips inside and put them into a rumbler to remove the soil. They'd then go into a slicer to produce food for the sheep and cattle."

Roger and Cynthia were both infants when the worst weather of the twentieth century hit in 1947. From 22 January until 17 March, records show that snow fell somewhere in Britain every single day, with temperatures rarely creeping over 2°C. In Halifax, where Cynthia lived, there were snow drifts seven feet high. What was particularly curious was the winter had been very mild up until the end of January, then nature did its absolute worst.

In 1962, it was a similar story when the first snows and bitter winds of The Big Freeze arrived just before Christmas. Temperatures plummeted, lakes and rivers froze over and most of Britain was covered in snow for the entire month of January. Even the sea froze in places and, in some rural areas of England, the weight of snow and ice stretched the copper telephone wires down to the ground.

Although the countryside couldn't have looked more beautiful when the sun shone, it was a challenging time for everyone who lived through it. "The winter of 1962 and 1963 was just hell," says Roger. "I remember driving Cynthia home when we were courting and having to abandon the car because it just wouldn't make it up the hill where she lived. The snow was virtually banking up the sides of our bodies as we battled through the drifts to her house. I remember that icy tickle as my ears filled with snow. When we got there, we couldn't even see her front door. It was completely buried, and we had to burrow our way in. Good God, it was cold up there. It was like stepping into the Arctic."

"Winters were so much harsher then," adds Cynthia. "Occasionally we'd have a white Christmas, but the snows always seemed to come later in January. I loved it though, it was so quiet and no cars to speak of. Because we lived on a hill, when we got the bus home from school, it would slow down and we'd have to jump off while it was moving. If it stopped, it would never have got moving again.

"And I remember one winter before Roger and I were married when part of the field was frozen," she says. "A sheep slipped down all the way into the river, so we planted a new hawthorn hedge to prevent it happening again. You couldn't see the grass at all; it was just ice everywhere."

3

Like the weather conditions, farming methods have changed dramatically over the last century – certainly where harvesting is concerned. Some farmers no longer need to grow the amount of crops and animal feed they once did, as specialist suppliers now produce everything a farm needs. Also, certain practices, such as threshing crops and grinding grain, wouldn't be practical on a farm that's open to the public. All that noise and dust wouldn't be appreciated by paying visitors, not to mention any sight of a fat brown rat – that would be a definite no-no.

With commercial feeds available, farmers like the Nicholsons have been able to rework what would have traditionally been arable farmland for different uses. Farming practices have also changed massively in the last century. Where once-heavy horses would be pulling the ploughs to work the land, now massive tractors do the hard work.

"I can still remember our two magnificent Shire horses Blossom and Captain, bringing in the harvested crops," says Roger. "They'd bring a cartload in, it would be unloaded, then they'd go and fetch another. It must have been just at the crossover stage with tractors, because after that I remember we had a tractor.

"Decades later, before the boys were born, we had two little

28

Fergies [Ferguson tractors] and they did everything that was needed. You could put a plough on them, or a cultivator or drill [seeder], and off you went. There wasn't a machine for this and something else for that. A good little tractor was all you needed. Even so, it was a major investment for the farm and a new one would have cost hundreds in those days. We always bought second-hand from a farm sale – not that we'd have to replace them that much though; they were built to last.

"I remember for years and years we had a David Brown 1390 with a loader on it," says David, the youngest of Roger and Cynthia's sons. "It wasn't exactly the Maserati of tractors or anything like that. Truth was, it was probably all we could afford at the time, but it was a real work horse."

From the basic tractor, farmers started using more sophisticated "combined" machines, hence the term combine harvester. They did several jobs in one, cutting the crop, lifting and threshing at the same time. "We started to use a combine in around the sixties, but bigger grain farms had probably been using them for many years by then," says Roger. "It completely replaced the threshing machine. The straw would come out of the back of the combine, the grain would be bagged up and the chaff would be scattered on to the land. Then the mechanical balers gathered the straw into bales. You'd hire a contractor who had a combine harvester, because a farmer wouldn't have enough acreage to justify having one of their own, and it would do everything. Now farms are so much bigger, they have their own combines as it's almost one hundred per cent industrialised."

While there are certainly advantages in the way that farmers can now work – a lot of the jobs that once spelled sheer drudgery are now automated – in terms of making a living and economies of scale, farming has become a very different proposition to what it once was. "When my dad first started out at

Cannon Hall Farm, two hundred and fifty acres was a big farm," says middle son Robert. "Of its time, it was a perfectly decent size, but with every decade that's passed, farmers have had to grow and grow. In the sixties, some people would milk twenty cows and be able to make a living, but by the time you got to the eighties, you'd have needed a hundred, then, by the end of the century, it was two hundred and fifty, and now it's more like four hundred. World commodity prices rule the roost."

Mechanisation slashed the costs of labour and is at its most efficient when working on large areas uninterrupted by barriers, such as hedging and drystone walling; that's why much of the landscape of farms has changed. Now, a single field may be same size as the whole of a traditional farm. And where there used to be mixed farms consisting of a variety of farm animals and various crops, modern farms will specialise in just one or the other – generally only producing a single type of crop or breed of animal for beef, dairy, lamb and so on.

Diversification has become almost as important as traditional farming to farmers, particularly those running smaller farms. "It became obvious around the mid-1980s that there was no future in running our farm on a purely agricultural basis," says Richard. "At Cannon Hall Farm we've always been interested in diversification and adding value to the business and we've won awards for it. We must now be one of the most diversified farms in the UK that retain a working farm at the heart of the business. Dad was always keen to retain the farm, as farming is his lifelong passion, and while there's no doubt we could have made more money and less work by diversifying the site completely away from the working farm, that would have been a huge failure to Dad. He never wanted to be the last Nicholson to farm and we're so pleased he's been able to create a successful business and make that possible. Everything we did was to keep the working farm viable and thriving, and we always felt we'd lose a lot of the integrity of our farm shop if we relied on other farms to supply us."

Over 50 per cent of UK farms now have an element of diversification. Tourism accommodation, renting business premises, countryside activities, catering in various forms, food manufacturing and farm visitor centres have all featured heavily in farm business diversifications in recent years. Unless you have a huge acreage, it's almost essential to survive. "We've been groundbreaking in our use of social media in the farming industry and there's no doubt it's revolutionised the way we market the farm attraction and associated businesses," Richard explains. "Being open-minded about the direction of the business and being on the front line of innovation has certainly made a big difference to our success going forward. We're really pleased that we've been able to keep it a family business and it's been great to see my son Marshall start working with the animals after leaving school. I'm hoping my dad's love of farming rubs off on him and the early signs are promising."

Changes in farming practices mean that farmers Robert and David no longer have to follow in Roger's footsteps in terms of growing crops for the animals, having to thresh grain and cut turnips and kale in the depths of freezing cold winter. Nevertheless, they do have the small matter of getting 100 ewes ready for lambing in February every year and a further 300 at Easter.

At Cannon Hall Farm, commercial sheep are kept, along with native favourites such as Herdwicks, Jacobs and Kerry Hills. The farm also has rare breeds of sheep like the gorgeous black-nosed Swiss Valais (think Shaun the Sheep) and the delicate-featured Soay. Twelve breeds are kept in all: some reared for commercial (meat) farming, such as the Charolais and Texel sheep; some will be sold to help strengthen the breeding lines; and others who will become farm favourites with the visitors. For example, Nemo, the little lamb born with one normal-sized ear and the other tiny. "We've always had a bit of a soft spot for animals that come out a bit wonky or underweight," says David. "It's a family thing. My dad's old sheepdog Flossie nearly didn't make it either but we often keep animals if they've had a tough start or they need extra help."

Ewes are pregnant for between 145 and 151 days, so therefore to guarantee that lambing coincides with the timing of the February half-term school holiday the first flock of around 100 ewes will have been mated the previous autumn. Farmers cannot confirm to the day when the ewes will lamb, but the system seems to work just fine. The sheep are then vaccinated the following January to prevent serious bacterial diseases including lamb dysentery and tetanus. Once their lambs have been born in February and grow big enough, they too will be vaccinated.

A team of around a dozen farmers work at Cannon Hall Farm, headed up by Robert, David and Roger – or, as he has come to be known, "Ranger Roger" – at the helm. That means that along with the trio of Nicholson farmers who live there, there's plenty of extra help every day to get everything done, from feeding and mucking out to exercising, grooming and seeing to medical needs. It also means that there are plenty of skilled farmers at hand during busy periods such as lambing time and when the television crews are on site when Robert and David are needed to film reports for the programmes.

Many of the staff have worked on the farm for several years, some for decades. Ask the majority of the team and they will tell you that what had started out as a bit of work experience/a little part-time job/a short-term contract turned into a full-time role, and that they now feel part of the family. They've been part of the Nicholson family's incredible farming journey every step of the way.

But not all farm jobs can be handled by the team based there. Fans of *On the Farm* will be familiar with Peter Wright, Julian

Norton, Shona Searson, Matt Smith and David Melleney who are often seen on the sister Channel 5 TV show *The Yorkshire Vet* (more about them at the end of this book). They visit Cannon Hall Farm for scheduled health checks and deal with emergencies on a regular basis.

One such emergency occurred in the middle of the night during January 2020 when David was on holiday and Robert and Roger were looking after the farm between them. One of their Shorthorn cows went into labour and they had a fight on their hands to save both the mother and her calf.

The previous year, to celebrate Roger's 60th year at Cannon Hall Farm, Richard, Robert and David had bought Roger a herd of 18 beautiful heifers and a bull called Jeremy. The Anniversary Herd, as it came to be known, was strong and healthy and no one could have envisaged the drama that was about to play out. To say there were teething troubles at the start would be putting it mildly. "We'd bought the herd thinking it would be a lovely present for Dad and that they'd bring him endless pleasure, but initially it was nothing but endless stress," says Robert. "Several of them were struck by a mystery virus and lost their calves. The weather conditions didn't help either; it was really hot for spring, then really cold at night and that may have triggered their illness. They got raging temperatures and that's what caused some of the calves to die inside their mothers and others to have difficulty giving birth.

"We scan the herd twice a year to find out how many are in calf, and to assess their due dates, so when it came round to January 2020, we knew one of the cows was heavily pregnant and would be calving any day," Robert explains. "She was really big and I could tell it was unlikely she would be able to give birth naturally and would therefore need a caesarean, so I called in the vet."

When an animal is as huge as a pregnant cow, the vet has to

come to the animal, not the other way around, so it's a field hospital situation. It's not like getting a lamb on to a trailer and zipping down the road to the vets. On a busy farm, the vet has to work in conditions that are far from ideal, as there's often a lot of muck and dirt, but they just have to do their best – which isn't easy when you've got a distraught cow bellowing at you. "You also have to be quite nippy on your feet because a cow can tread on you or shove you if they're feeling stressed," says Robert. "If a medical emergency happens during the daytime, it's easier for everyone as it's light and you're more likely to have plenty of people around you to help, but on this occasion it really was the worst-case scenario – just me and my dad in the middle of the night – in the middle of winter.

"Matt Smith, the vet that came that night, did a wonderful job getting the calf out of the side door [the term for a caesarean] but the calf didn't seem big enough for the sheer size of its mum. It had only been scanned for one calf but I said to Matt, 'I don't suppose there's another one in there?' and sure enough there was. But, when he brought the calf out, we could see immediately that the second one hadn't formed correctly. It had birth defects and couldn't bend its legs properly. Sadly it didn't survive, but the first calf grew up healthy and strong, its mum recovered well and she was able to get back in calf again the following year."

Head of the Anniversary Herd, Jeremy the bull was in the spotlight in early 2021 when the herd was once again featured on *Springtime on the Farm*. Although he looked happy enough to be with his bovine harem, David spotted his foot was giving him a bit of gyp. A specialist was contacted to trim away his hooves and an infection was identified. He was helped in the nick of time. "All it takes is for an animal to get a little cut and if it gets infected it can get worse very quickly," says David.

"Unfortunately for Jeremy – and for us – he's a one-bull slurry machine, so it's not that easy to keep him clean."

Jeremy and his ladies are the first herd of Shorthorn cows that have been on the farm for many years. When Roger was growing up, they kept mostly dairy cows, but his father always favoured Shorthorns as they were seen as a good "dual-purpose" cow – meaning that they were reared for both beef and milk. "The thing is, those dual-purpose cows were all right for beef and all right for milk, but they weren't really right good at either," says Roger.

Shorthorns can be red, white and roan, which is a marbled mixture of the two colours, and whereas once there was just a single breed, now there are two distinct breeds of Shorthorn – one bred for dairy and one bred for beef, which is the variety of the Anniversary Herd. "The two breeds have the same colour characteristics," says Roger, "but their physiques are very different. A dairy Shorthorn is very light-boned with not much flesh on it, whereas a beef Shorthorn is more robust and has a better conformation."

Around 90 per cent of Britain's cows are Holstein-Friesian, the distinctive black and white cows from which we get the majority of our milk, with Ayrshire, Jersey and Guernsey being the next most popular with dairy farmers. For beef production, the most common breeds reared in the UK are Limousin, Aberdeen Angus and the Charolais. "It's totally up to the individual farmer which breeds they'll choose for their farm," says Roger. "First, it's about what you want to get from the cow – are you breeding them commercially for beef or dairy? In which case you need to think about which breed will thrive best on your land. The old-fashioned Shorthorn used to be the main breed in Yorkshire and our Highland cattle do very well in Yorkshire because they are used to the hilly terrain."

The same goes for sheep – farmers decide which breed they

want to rear on their land, whether they are rearing sheep commercially or to sell on as breeding stock. There are over 60 breeds of sheep in the UK, more than any other country in the world, and although they share many of the same characteristics, they vary greatly in the looks, size and shape department.

"Most breeds thrive in the conditions we have in Yorkshire," says Roger. "Native Yorkshire breeds are mostly upland sheep so they are quite hardy, and some breeds much prefer lambing out in the fields than being indoors. Certain breeds like the Herdwicks can get itchy as they are more susceptible to getting a mite on them when they are inside and that makes them very uncomfortable."

In the old days of farming, sheep would be dipped in a trough of strong chemicals, which would kill any bugs in their fleece, but now they are given injections that kill internal and external parasites. "There are so many rules and regulations about the disposal of sheep dip that it's generally not used any more," says Roger. "It used to be very effective in stopping the spread of infection as the chemicals would remain in the sheep's fleece so that, once they were dipped, they were protected."

In many farming practices, methods have evolved which are healthier for both livestock and the environment. Each breed's specific needs are considered and farmers will have an agenda of annual health checks and vaccinations to ensure their animals stay happy and healthy. Fortunately, there's no need to call out a vet to administer most farm animal vaccines and the team at Cannon Hall Farm are very nifty when it comes to quickly and carefully vaccinating their herds to keep them safe.

"Vaccinations have always been done by the farmer in my lifetime," says Robert. "It's something that I learned to do at college. I can see a day when farmers are doing a lot more because there just aren't the numbers of large animal vets that there used to be. We are really lucky here because we have local vets Matt, Shona and David and they are the real warriors and all-rounders as they will treat all animals, large or small."

Many veterinary practices are cutting down on their farm animal workload and will only treat pets and more exotic animals in the surgery because it just isn't cost-effective to travel miles across the country to treat one farm animal. "I think farmers will be trained to do more," says Robert. "Not surgery and caesareans and things like that, just more everyday procedures. While I could learn to do more, I'm not desperate to, to be honest. There's a reason why they are vets and I'm just a farmer!"

When there's a specific job that needs to be done on the farm, there's a specialist to do it. One such necessary practice is scanning the pregnant ewes early in the year to find out how many lambs they are carrying. The scanning operator visits the farm when the ewes are halfway through their pregnancy – roughly around 70 days after the tup (ram) has mated with them. The ewes are rounded up and brought together to be scanned using a similar ultrasound technique to the ones used in human birthing – albeit with a little less finesse than a human mum might experience. Scanning identifies whether a ewe is pregnant with a single lamb, twins, triplets or, in some cases, quads. In extremely rare cases, a ewe can give birth to quintuplets, but to date this has never happened at Cannon Hall Farm.

"Scanning is one of the most important jobs of the year," says David. "In a perfect world, we'll get a pair of lambs to each mum as a ewe has two teats, which means that each lamb gets its fair share of its mum's milk. We usually get about eighty single lambs a year and the same number of triplets," he adds. "Having the third lamb (or more) obviously causes problems as it will either have to be adopted on to a ewe who's only had a single lamb, or it will need to be bottle-fed."

In 2020, during the two lambing periods, 800 lambs were produced from the farm's herd of 450 ewes. In an episode of *Springtime on the Farm*, pregnant ewes were scanned and marked with a spray of dye to indicate how many lambs they were carrying. In the case of the Nicholsons' sheep, a mark made on the shoulder means a single lamb, a mark on the bottom indicates triplets and a mark in the middle of their backs signifies no lambs are present. For twins, there's no need to make a mark as the expected number will always be two. When it's quads, it's a double spray and, in the unlikely event of quintuplets, the whole paintbox of colours comes out!

Marking ewes this way is a basic technique that has been used in farming for decades, but it's an easy and clear system and also means that the farmer knows just how much each ewe needs to be fed, depending on the number of lambs she is carrying. The ewe's nutritional needs are then balanced so they will be nourished enough to produce strong, healthy young. Handled correctly and the lambs will be a good size, but won't be too big so that the ewes will have problems delivering them.

With the first of the spring lambs scanned, Robert and David can then start any supplementary feeding that's required. The ewes are rounded up into pens depending on whether they are carrying singles, twins, etc., and fed accordingly. Single lambers get the usual amount of hay, those with twins get a boost of extra grain, and ewes having triplets and quads are also fed a quickly absorbed sugar-based substance containing molasses. "The big danger of giving ewes a lot of food is that the volume in their tummy can increase so much it can cause the uterus to prolapse during labour," says Rob.

One such moment happened in an early episode of *Springtime on the Farm* in 2018. Luckily, vet Julian Norton happened to be on hand as he was appearing in a later item in the show. Luckier still, he'd brought his bag of veterinary equipment with him. The ewe had strained so forcefully while it was giving birth that it had pushed the lining of its uterus out of its body. But, with careful manipulation, Julian was able to ease the womb back into place. His intervention was so successful that, the following spring, the lucky sheep was able to carry another set of lambs.

Over at Mill Farm, it was Richard to the rescue when he discovered a shearling in distress in 2020. "Clare [Richard's partner] heard a desperate bleating call from a field near where we live," says Richard. "I set off to find the source of the cries and there, in the middle of the field in a pool of blood, lay an exhausted ewe who'd given birth and prolapsed. A group of opportunist magpies were pecking away at her rear end, causing more damage with their dagger-like beaks."

Luckily, Robert was visiting his daughter Katie who lives nearby and he was able to quickly lift the ewe into the back of the Land Rover. "He then managed to corner the lamb with a diving catch he'd have been proud of in his cricketing days," says Richard. Back at the farm, Robert and David worked quickly, washing the uterus to try to prevent infection, pushing it back inside the ewe, making sure everything was positioned correctly, and fitting a restrainer so it couldn't push its womb back out again. A quick dose of antibiotics and painkiller and fingers were firmly crossed. "Her condition when we found her was so bad we hardly dared hope she would make it," says Richard, "but she was determined to rear her lamb. Robert described her as 'nothing short of heroic'." Incredibly, the plucky young mum made a full recovery and even went on to deliver and rear another lamb without any problems the following lambing season.

Hundreds of sheep have been born at Cannon Hall Farm, and there will be countless more in years to come, so it's vital that each one of them has the very best care and attention they need every day of their lives. From the practical side, farmers need to make their business work, so, in all cases, they strive for optimum health. Whether an animal is being reared as a breeding stock, as an addition to the open farm's animal collection or as high-quality produce for the farm shop, every aspect of its care is important. That means not only its health, but caring for the environment it lives in.

As well as nurturing their flocks of sheep, the family have always utilised all of the farm's available space. Before Cannon Hall Farm was expanded and opened to the public in 1989, the Nicholson family used each of their farm buildings for different purposes, depending on their requirements at the particular time of year. At Christmastime, for example, once the turkeys had all been sold off, the barn where they had been kept could be swept clean and made ready for the next cereal crop or animal it would be housing. Even today, with all of the farm's purpose-built facilities for each and every breed, Roger still uses the original outbuildings (also called boxes) to house everything from well-used farming equipment, or a clutch of duck eggs being carefully incubated, to his five white Leghorn laying hens that provide fresh eggs for the family every day. All of the animals' barns, tanks and enclosures are carefully maintained to create the optimum environment for each particular breed to thrive.

4

Making sure animals have the optimum conditions to flourish means keeping check on the boundaries that contain them, like the acres of hedgerows and drystone walling in South Yorkshire that stitch the fields together like a rural patchwork quilt. When you look out at the view of the countryside, it's these markers that indicate the extent of the landscape, drawing the eye to the different colours and textures of the crops and the areas where animals are grazing. Whether they are borders of bracken and wild roses, hawthorn hedges or expertly built drystone walls.

In Britain, there are so many different styles of drystone walling that it's sometimes possible to be able to identify a county simply by looking at the handmade stone field boundaries. For example, in southern and central England, walls are built from russet-brown and golden-grey limestone, and in areas of Cornwall, walls can be found in a distinctive herringbone pattern.

Drystone walling dates back to the Iron Age and, while many of the stone walls at Cannon Hall Farm aren't quite that old, they would have originally been built several centuries ago to act as land boundaries between properties. In the Yorkshire

Dales National Park alone, it's estimated that there are around 5,000 miles of drystone walls. They're built from different sized pieces of stone that are found locally, often thrown up during ploughing. They're collected, then expertly pieced together in layers – the magic being that no mortar or cement is used to hold them in place – hence the term "drystone". These walls are incredibly resilient and generally need little maintenance. Over time, trees and some shrubs can destabilise and uproot the structure of a wall and extreme weather can cause top stones to topple, but it would take a pretty deter-mined force to be able to knock down a well-built wall completely. You'd expect a good drystone wall to last 100 years.

In one episode of *Springtime on the Farm*, viewers enjoyed watching Robert and David as they tried their hand at the ancient skill, repairing a wall that had partially collapsed on their land. With lambing season about to start, drystone walls provide shelter for the lambs born out in the fields. The walls act as a windbreak from the bitter northerly gales and would also provide a firebreak if necessary, plus they're a great habitat for hundreds of insects and mini-beasts, mosses and wildflow-ers. The really amazing thing about these handmade walls is that they actually become stronger and more closely bound with time. And, as drystone walls use only natural material, they're one of the earliest examples of recycling.

If Robert and David had been building a drystone wall from scratch, it would have been necessary to prepare the ground, lay foundations and gradually build up the layers, making sure to keep the wall stable. They would then top the wall with flat stones called top, cap or coping stones. It sounds straightfor-ward enough, but whenever Robert and David are trying their hand at a new skill, their competitive nature has to be factored in to the proceedings as the brothers constantly like to outdo each other. And this episode of *Springtime on the Farm* was no different – the two brothers saw the comedic value in working on opposite sides of the wall, trying to steal each other's best stones to get the best results. In the end they asked the drystone walling expert, Joe Smith, to choose who, out of the two of them, had done the best job. Luckily, he proclaimed it a tie.

Roger has much more hands-on experience than his sons in the art of drystone walling. "I helped build some of the walls from scratch when we converted the Mill Farm property in the Gunthwaite Valley," he says. "It's something I've done all my life. Like most things, I watched an expert do it first, and that's how I learned as I didn't have the benefit of college or owt like that. It's just a matter of common sense, really – using the best stones on the outside and smaller packing stones in the middle. You build your wall up with a slight camber, so the whole structure is supported from the base up.

"Walling is a wonderful thing to be able to do as it can feel very therapeutic to see all the layers building up. You can take real pride in your work. But I've known lots of people who have had to give it up because of the strain on the fingers. You need powerful grip strength to be able to constantly lift up heavy pieces of stone using just a small part of your hand."

As nature does its work throughout January, farmers like the Nicholsons have to keep constant checks that perimeters have stayed in position, that electric fences are still operating, trees haven't become uprooted and that banks of rivers and streams haven't been eroded by flooding. No matter how much you check, it only takes one wild storm to cause incredible damage.

Electric fences are positioned around parts of the farm to keep animals from straying. And, although they sound a little scary, they have a low pulsating current, so they are not dangerous to humans or animals. Saying that, if you come into contact with one, you'll certainly know about it. Roger once discovered this to his cost . . . "I was looking around for my sheep one day and I nudged my head over a wall into a neighbour's garden – not realising there was an electric fence there. I put my chin on it and it nearly locked my teeth together! That's what comes of being nosey!"

—

Another tradition that still survives to this day is spring cleaning, be it on a farm or any other dwelling. Maybe once the new year's resolutions have been ticked off the list, the next thing is to have a good old clear-out – or perhaps it's just because on a nice bright day you can see it's time for a deep clean? Whatever the reason, there's nothing like getting your Marigolds on and giving the house a good spruce up. "My mother kept a very clean house all-year round," says Roger. "She was forever going round with the carpet sweeper and she'd have my sisters out beating the rugs once a week. But, when it came to spring, she was like a thing possessed. I'd scarper sharpish while she attacked the house like a whirling dervish. The heavy curtains would be down, mattresses turned and she'd use a rags, bicarbonate of soda and white vinegar to give everything an intensive clean. I don't think she was ever happier!

"I remember one time I went to her house and she was bent down, scrubbing the hall floor," says Robert. "I didn't see her through the glass in the door and I threw the door open, straight into her head and nearly knocked her out."

"That's where cleaning gets you," says David. "Don't take any chances!"

Always one to be philosophical about things, Robert sees spring differently: "For me, the start of the year is the time when we can take stock of everything that is going on and what we need to do to make things happen the way we'd like them to throughout the spring and the rest of the year. January is very much a maintenance month. That means repairing anything that's had a battering by the weather and making sure that we have all that we need in terms of vaccination supplies and everything ready for the first lambing of the year.

"It's also very much a period of both reflection and moving forwards. That means looking back on the things that we could have done better in the previous year and learning from our mistakes and also feeling positive about what's to come. If we've been doing our job right and we're well prepared, then hopefully, we won't be under too much pressure for the first really busy period of spring when we get our initial big influx of visitors."

It's a very different story to the time when Richard, Robert and David were little and didn't need to plan anything from day to day. In January most years, they would be more preoccupied with knowing what was for tea, or praying for early snowfall so they could have a day off school. "I remember the three of us would wake up various times during the night and peek through the curtains to see if it had snowed enough to stop us going to school," says Robert. "There was a great deal of excitement if that happened, but we'd usually have to wait till February until we got a good covering. Until then, January used to be a bit of a dead month. I suppose we should have been grateful for having nowt to do because, as soon as we were old enough, Dad would have the three of us out picking turnips in the worst of winter. It was such a bloody awful job!

"Our biggest nemesis would be trying to keep on top of the pipes freezing," he adds. "We had one outside tap for the whole of the farm and if we had a really solid spell of ice in January, we'd be stood there with a blow lamp trying to get the pipes to thaw, so that we could fill the animals' water troughs. But gas was expensive so the blow lamp would have to be our last resort.

More often than not a frozen pipe would mean carting water from the farmhouse in tubs, buckets and anything we could lay our hands on, and sloshing it into the water troughs. You could find yourself spending a whole day doing that – going from one freezing cold barn to the next and then back round again. It was such a dispiriting time, but water is fundamental to an animal's life; if it can't drink, it can't survive."

Thankfully, life for the boys wasn't all about doing chores, and they were always happier to help out than to see their dad struggle. It's always been a team effort. "Richard, David and I were all so lucky in that respect," says Robert. "I never remember having to do a whole list of chores before we could go out and play, and Mum and Dad never forced us to go into farming. They would have supported us no matter what we decided to do. They just wanted us to be happy."

Sometimes the early months of the year could be quite a drudge for the three brothers when the rain never seemed to stop and it was just too horrible to go outside. "It never seemed fair to drag the kids round with us when we were feeding and mucking out the animals," says Cynthia. "And TV was very different back then. Only three channels and nothing much for kids to watch – apart from *Watch with Mother* and an hour of children's TV every afternoon.

"I have to say they were pretty good, though. I'd keep old rolls of wallpaper for Richard because he always loved art and he'd draw life-sized superheroes on the back of them."

"I'd spend hours working on them, with the paper stretched out over the floor of the front room," says Richard. "Spider-Man, the Hulk, Captain America . . . Every Friday, Mum would go shopping with her friend Rosemary and return with a comic for each of us; *Spider-Man* magazine for me, *The Beano* for Robert, with his hero, Dennis the Menace, and *Nutty* for David, with his favourite, Bananaman. I got my first

Spider-Man comic when I was about eight and it started a life-long love of Marvel superheroes. I still enjoy watching the films with my son Marshall now."

Keeping pets was also a big hobby for the boys. "At first, they started off with the odd hen," says Cynthia, "and when they got older, they had snakes and lizards and all sorts in a tank in their room. And yes, they did escape sometimes – poor grandma Olive never forgave them for the day she awoke with a coiled snake on her head. Apart from that it was the usual board games like Ludo and Snakes and Ladders. They were very competitive – even back then. They also played football in the house with the doorways for goalposts. I didn't mind though; it wasn't as if we had any fancy china or ornaments to break anyway. And, to be honest, I'd often join in!"

"We also had a rather weird hobby of finger knitting," says Richard. "It was featured on *Blue Peter* and was a mindless and pointless activity that created a string of wool by knitting wool from one strand into three. We tried it and soon we were all at it, knitting away furiously for no apparent reason. We ended up with a ball of finger knitting big enough to sit on. Heaven knows why we liked doing it but it seemed somehow therapeutic at the time."

Living on a farm meant Richard, Robert and David always had loads of space to play, just like their dad had when he was a little boy. "We had lots of friends round all the time," says Robert. "My mate Darren was always coming over to spend days at the farm. He was a bit naughty, but he was a real character and we'd spend hours and hours playing football together. When our Dutch barn was full of hay, we used to play football outside or in the house, but as it got emptier and emptier it would become our football pitch. It wasn't the size of Wembley, but it was certainly big enough for a good game of five-a-side. We'd chalk goalposts on the walls at either end of the barn and it would be Leeds United v Barnsley FC. In our imaginations we'd be playing in the European Cup Final and, depending how I felt, I'd be Gordon McQueen, Joe Jordan or Johnny Giles. Meanwhile, Darren did his best Jimmy Hill impression and would commentate on our skills."

The old hay barn is now where the award-winning Cannon Hall Farm shop stands. It still has the original trusses where young Robert and Darren would make rope swings for death-defying aerobatics. "I remember that I'd be constantly looking up and checking the fraying rope wasn't about to break – otherwise we would have been in peril."

There are now much more sophisticated play areas for children at Cannon Hall Farm – both indoors and outdoors. This has meant that when Richard, Robert and David went on to have children of their own, their four had their own farm to play in. Every day, before going to school and at night when the public had gone home, Tom, Katie, Poppy and Marshall could try out every new swing, climbing frame and slide for themselves. They also had a few hundred farm animals to enjoy whenever they liked.

Tom, Katie and Poppy have all grown up and left their childhood home now, but Richard's son Marshall still lives at Mill Farm. He provides an extra pair of farmer's eyes for Roger, keeping check on all the animals that graze there. He's saved several sheep that have been caught up in fencing or taken a dip in the millpond. Meanwhile, Katie's little daughter Nelly also shows she's a chip off the Nicholson block as even the biggest animals on the farm don't seem to faze her. Maybe she will also follow in the family footsteps?

5

Every year, provided Robert and David have been given the go-ahead with a new series of *Springtime on the Farm* or *This Week on the Farm*, January is a good time to get some filming in the bag. The two brothers have a bit more time on their hands compared to frantic February, when lambing kicks in, so they don their thinking caps for ideas that could make good stories for the programme. Maybe a follow-up on one of the farm favourites – an expectant llama perhaps? – or a road trip to buy a rare breed for the farm.

"We're usually told by Christmas whether there will be a new series the following spring," says Robert. "So we start filming reports as soon as we can. That way when it comes to the live programmes going out, we can concentrate on lambing. In spring 2021, we filmed right through January, February and March, with six episodes of *This Week on the Farm* and straight into four episodes of *Springtime on the Farm*. So, as well as all of the live shows, we had various other films to make in advance. It was hectic, to say the least."

Regardless of the season, every month there are new births on the farm, whether it be a scheduled litter of piglets, or an unexpected birth – for example, in summer 2020, an escape artist billy goat managed to check in with one of the ladies. (Robert and David still argue about who left the gate open that time!) Plus there are always the day-to-day farming jobs to be carried out: pens, paddocks, barns, tanks and exotic habitats all have to be maintained; there are health checks and other husbandry to be kept on top of; and there's always a mountain of administration that needs to be dealt with in order to run a successful working farm.

For the staff who do these jobs again and again, year in year out, many elements of farming may seem completely routine and not exactly the stuff of unmissable viewing, but it's pure "edutainment" for fans of *On the Farm* and Cannon Hall Farm's online followers. During the 2020–21 lockdown, the feedback that Robert and David received from viewers was really touching, with many saying their programmes provided pure escapism from the stresses of Covid-19. Seeing something as fundamental to nature as a lamb being born was a breath of fresh air in a climate of confusion. You could switch off the real world and escape to the farm, meet friends old and new and immerse yourself in the stunning Yorkshire countryside.

Cannon Hall Farm supporters can also access reports all year round via the farm's official website, often thanks to farmer and resident Steven-Spielberg-wannabe, aka Dale Lavender. Dale first came to work for Cannon Hall Farm in 2007. While he was on holiday one year, he learned how to use a video camera and he started making short film reports about the animals at Cannon Hall Farm that were shown on the farm's website. In fact, Dale became such an expert behind the camera that, during the 2020 lockdown period, he captured several reports that were of professional enough quality to be shown on *Springtime on the Farm* and *This Week on the Farm*. "When the TV company wasn't able to access the farm, they'd send me a script of the shots that they would like," Dale explains. "It could be a wide shot of this, or a close-up of Robert and David talking, that kind of thing. Sometimes I'd think, 'This really isn't going to be any good,' but when I saw it back on TV I was quite pleased with it. The editors obviously know how to make something very amateur look very professional!"

Often with Dale at their side, all three Nicholson brothers are so adept at filming for social media, they can take full advantage of any quiet times to get ahead with filming items for TV. "David and I come up with some ideas and suggest them to the production team, who work up the items from there," Robert says. "I'm yet to convince them that we should go on farming road trips all around the world, but I'm working on it!"

Once an idea for a story has been agreed on, the production team create a shooting script, which is an outline of how the story needs to look to convey it to television viewers. It covers all manner of logistical hurdling blocks that may have to be crossed (such as permission to film at certain locations) and will include lists of scenes and camera angles that the director hopes to capture. It also includes the dialogue script for those in front of the camera to follow. However, anyone who has seen any of Robert and David's programmes knows that the brothers don't always stick to the script . . . "We have licence to adapt and vary it," says David. "If something happens that makes it funnier, or we are able to put our own twist on it, the programme makers like us to go for it. I think the best bits are when we are allowed to just react to whatever happens. I also like to chip in with ideas about how we make films sometimes, so I've earned the nickname Director Dave. For some reason, I haven't got my own director's chair though . . ."

For each report, the camera will need to film each shot a number of times to include different angles, so that the scene is seen from different points of view. That means repeating the dialogue again and again – and trying to stick to the script. "If we mess up, we have to do the shot again," says David. "We do try and nail it the first time, but it's not unusual for me to crack up laughing. One time we were filming a report about Busty the alpaca and it was really late in the day. I was just about to say my line and Jules Hudson [one of the show's presenters] made me burst out laughing. After that I got the giggles and I just couldn't stop. All the crew had very long faces because they just wanted to finish up and go home. Meanwhile, I was trying desperately hard to compose myself."

On an average day, depending on whether they are out on location or filming at the farm, Robert and David, or whoever else in the family is needed, will be at the location for 9am along with the production team. Everyone will get cracking filming, there's a quick break for lunch, then it's back to work until everyone's happy that all the necessary shots are done and dusted. But while the day's work may be over for the TV production team, it's back to the day job for Robert and David – especially in the middle of lambing time.

All this is in preparation for the live week of *Springtime on the Farm* series, which airs every April. This means that once Robert and David literally have their hands full with lambing while the show is on live, there will be a wide variety of other stories that they will already have filmed that are ready to roll during the show.

For the programme makers, keeping the shows fresh and entertaining is all about finding a mix of traditional farming stories and filming enlightening, innovative reports on a whole spectrum of rural subjects. Ancient crafts are celebrated alongside modern methods of producing food and rearing animals. There's the serious side, when an animal loses its fight for life or a farmer has to call time on their business, along with plenty of upbeat tales from Cannon Hall Farm.

"Creating the programme is like putting together a huge

jigsaw," says Daisybeck's Paul Stead. "We look at the shape of the show and the shape of the season, the ideas that Rob and Dave have focusing on Cannon Hall Farm and the reports that JB Gill will film for us, plus other interesting stories from around the country. There's a senior editorial team of four people from Daisybeck and Channel 5 who decide on the balance of each programme, and a team of producers and researchers who will find the stories and pitch the ideas to us."

"There are some items we've filmed that have never been used," says Robert. "One time we spent ages making owl nesting boxes and putting them up, only for a cat to occupy one of them, and a jackdaw to move into the other! We keep checking in on the nesting boxes though and, you never know, the rightful owners may move in one day, so that film may not be wasted after all."

PART TWO
February

6

It's the shortest month, the month of Valentine's Day and Pancake Tuesday, but February is especially significant at Cannon Hall Farm as it's when spring really starts to kick into action. Beautiful bouncing, bleating lambs will be born in abundance, and the first of the year's flowers start to appear. Snowdrops may already be using all their might to push through the frozen ground even though the crocuses may not yet feel brave enough to put in an appearance. It's still too early for the migratory birds to return from warmer climes, but the blackbirds and robins are in full voice. "For me, the song of the blackbird is the anthem of spring," says Robert. "The thing I look forward to most in the mornings is walking into the farmyard and hearing the blackbirds singing, even when it's pitch-black outside and there's still snow on the ground."

February is also the month when it's most likely to snow in South Yorkshire. For some people this would be a pain in the neck, but not the Nicholsons. In a 2021 episode of *Springtime on the Farm*, Robert and David took to the snow-covered Buttercup Hill at Mill Farm along with their trusty sledge. "Old Faithful" was originally owned by Roger's father Charlie and was a Flexible Flyer, a design classic from America. Sadly, on the day of filming, conditions weren't ideal for the vintage shape of the Flexible Flyer – perhaps it was the wrong kind of snow? So Robert and David had to switch to more modern, streamlined sledges to conquer the powdery piste. They may both be in their fifties, but Robert and Dave were instantly just big kids again and their snowy shenanigans brought back wonderful memories.

"When we had a snow day and had to stay off school, it was just the best," says Robert. "We'd take Old Faithful to the steepest part of the driveway and shine the roads up so that we could get a really fast track. The head gardener at Cannon Hall Park used to try and stop us making our own version of the Cresta Run, but we'd just wait till he wasn't looking then get back to shining the ice. Then we'd be off, bombing down the slope, holding on to the Flexible Flyer and bailing out just before we hit the wall. You needed a parachute to slow it down, but miraculously, none of us ever broke any bones. We were fearless."

"I remember building snowmen and snow forts until my fingers were numb with the cold," says Richard, "and snowball fights that ended with that heart-stopping moment when snow was shoved down your back. Clothes didn't seem to keep out the cold very well back then but we didn't care. In the end, we'd tramp indoors, wet through, freezing yet sweaty, rosy cheeked from our play, feet like blocks of ice. Mum loved the snow too and would always insist on having a go on the sledge with us – even when she became a grandma. When she saw it snowing she'd say 'Ally ally aster! Snow snow faster!', which is something I've never heard anyone else say, but she'd always say it as soon as the snow began to fall."

Roger was always a snow nut too. "Our old school playground in Worsbrough Dale was on a slope so it was perfect for sliding," he says. "We'd all go out together and pack down the snow, then slide down it over and over again until it was a lethal sheet of ice.

By the end of the day, it would be so slippery it was like wet glass. It really hurt if you fell down on it, but teachers seemed to take more of a risk in those days. They'd never let you do that now, of course. God knows how we didn't break every bone in our bodies; I suppose people drank more milk and we had more calcium-rich food in our diets back then. We lived on milk and eggs."

"We'd be out there all day if we could when it was snowing," says Richard, "and we'd have huge snowball fights at school. We'd meet on Redgra, which, when not covered in snow, was a large, hard, red-surfaced sports field that would leave you with a serious gravel rash if you fell over on it. For the school snow-ball fights, it would be sixth form against the rest of the school – potentially me and the rest of the sixth formers against Rob and Dave and all their school mates. We'd start edging closer to each other with armfuls of snowballs, then all hell would break loose as battle commenced. It could get quite hairy if you got caught by the other side! You'd get shoved face down into the snow and have snow kicked all over you. Not great if you were on the losing side. The school we were at – Penistone Grammar – was certainly a school of hard knocks and the teachers mostly let us get on with it."

"Whether they were sledging on Old Faithful, trying to build an igloo or a snowman or pelting each other with snow-balls, the boys loved snow – and I did too!" says Cynthia. "The boys could keep themselves entertained for days when we got a proper covering – which is quite a relief when you've got three bundles of energy like them."

—

As fun as it may be to play in the snow, it's a different matter when you have to work in it – especially where farming is concerned. Farmers might like the diversion of a snowball fight as much as the next person, but it's no fun for anyone having to battle snowdrifts to stop your animals from freezing to death. Snow can be your worst nightmare when it's minus 15°C and there's a deafening, bone-numbing wind whipping the drifts up around you, and your head torch barely cuts through the darkness. But when you have a ewe out there and you know it's in trouble, you'll always do your best for it.

Several of the hardier breeds of sheep in the UK will always lamb out in the fields – rain, snow or shine – but at Cannon Hall Farm, the vast majority of their flock is ferried inside to lamb in the Roundhouse, their indoor-outdoor barn.

The Roundhouse was built on the farm in 2009 and its unique design allows easy entry and exit from each of the pens. It's open on all sides but covered by a roof so while it's sheltered from the elements, air can still circulate around inside it, which is vital in helping to prevent the spread of infection and disease. Roger was all in favour of the new innovation when the idea was first mooted. "I'd read about them and heard that a barn could be constructed a certain way with moveable metal sheets, and that the weather never gets to the sheep so they're always nice and warm," he says. "Sheep can stand being wet outside to a certain degree, but they're difficult to handle when they are sodden, and wet bedding is a real problem because disease will easily spread."

Lambing inside means the farmers also have easy access to any emergency supplies that might be needed, rather than having to carry lots of equipment around with them or bring an ailing sheep back from the fields. Knowing that you'll get a phone signal is also reassuring in the event of needing to call out a vet. Out in the wilds of a remote farm, this isn't always a given.

"Because lambing used to start much later in spring than it does now, the sheep would generally be outside," Roger explains.

"It was only when there were life-threatening conditions that you'd bring them inside; otherwise you'd leave them be because they're equipped to stand most of the weather," he adds. "Night-time was always the biggest worry though because you can't watch them twenty-four hours a day when they're out in the fields. And, if there's a problem, it's generally too late by the time you reach them."

In February 2021, farmer Dale was checking the sheep in the fields when he found a ewe stuck on her back. Her fleece was sodden and she was very shaken up. In a film he made for the farm's website, he explained that it had been raining solidly for around 12 hours and there was no telling how long the poor ewe had been stuck on her back. He had to act quickly. Back at the farm, it took two farmers to lift the waterlogged ewe out of the rescue truck and into a pen, then farmer Ruth gave her a good check-over. "If she's been stuck for a long time, she's going to be really run down and won't know what's going on," Ruth explained, "so we'll treat her for any infections and give her antibiotics and pain medication so that we can give her every chance of recovery. We'll also give her some molasses, which is full of the energy she needs to help build her up as well."

After a few hours of drying off and lots of food and water, the ewe was soon looking better and the farming team were delighted that she had made a full recovery. But although it would have been tempting to keep her in the Roundhouse, it was important to get her back to the fields. "The worst thing you can do for sheep and lambs is to keep them inside for too long," says Roger. "It's a bit like people being in hospital – they'll pick up bugs. Sheep are better placed than other animals to cope with infections outside, although in the worst of weathers, you can incur losses too. It's a case of working out what will be the least worst option. You need to assess how hardy the breed is and constantly work with the weather."

As lambing time approaches, the farmers gear up for what they know will be a very busy period for everyone involved. There can be some premature births as well as late ones, but the important thing is that everything is in place in order for them to hit the ground running. "It's no good flapping around getting your lambing sundries when lambing's started," says Roger. "It needs to all be in place beforehand." That means co-ordinating a shopping list of syringes, needles, antibiotics, pain relief and iodine and other medications as well as feeding equipment, heat lamps, foot trimmers and shears to cover any possible eventuality.

Like many animals, sheep prefer to give birth at night-time, so each day before the late shift begins, the farmers construct small square pens for the new arrivals and their mums – like a little animal hotel waiting for new check-ins. "We'll always try and let the ewes birth naturally, with no assistance from us," says David. "But we are there for when and if there are any problems. Once we see that their lambing has started, we leave her around forty-five minutes to an hour; then, if the lamb hasn't yet appeared, or we can see the ewe is struggling, we step in to help."

Regular viewers of *Springtime on the Farm* will have seen the brothers in action on numerous occasions during lambing time, carefully assessing what needs to be done when a ewe is in distress. Whoever is on shift will be constantly checking all of the ewes that are due to lamb for signs that it's about to happen. The ewe might take herself off to a quiet corner, start pawing the ground, sniffing around, pulling her head back and moving around restlessly – all signs that lambing is about to begin. When her waters break, contractions quicken and, if her womb is dilated enough, then it's all systems go. The whole process can be as quick as 15–20 minutes, but the birth itself usually takes around half an hour, unless it's a particularly big lamb on the way or there are birthing complications.

The textbook way for a lamb to enter the world is with its head first and front legs underneath its chin, almost as if it's doing a forward dive. But it's not always so straightforward, and that's where the farmers step in. The lamb may be in a strange position – upside down, back legs first or twisted up with other unborn lambs. It can be quite a jumble and without assistance, the lamb and its mum are very likely to be in danger. "There are lots of different techniques that we use to assist lambing," says David. "The main one is to put your hands inside the ewe's womb, position the legs forward, grip them in-between your fingers and gently pull the lamb downwards."

Sometimes lambing ropes are needed to get more of a grip on the unborn lamb, tying them around its legs to get more purchase before pulling it out. In some cases, a second rope is positioned behind the lamb's ears to ensure the head is in the right position, before gently steering the lamb out of the ewe's womb. "The skill is getting the rope into the mum and carefully around the lamb," says David. "You don't want to start pulling until you know for definite that you can get the lamb out safely, because if the umbilical cord breaks when the lamb is still inside, it will breathe in the birth fluids and you could lose it. You have to stay very calm."

Since the first days when local news teams came to Cannon Hall Farm to film sessions of lambing live on TV, Robert and David are used to having an audience watch them as they work. Because they schedule the first batch of lambing to coincide with schools' February half-term holidays, they know there will often be visitors watching them in action. They also have several series of *Springtime on the Farm* under their belts, so being surrounded by curious onlookers doesn't fluster them.

"Sometimes I just close my eyes, zone out of everything going on around me and concentrate for a few seconds on what I need to do," says David. "Nine out of ten times, it will just be a matter of gently easing out the lamb and positioning it near to its mum's mouth and letting the natural bonding process begin. However, there have been times when I have got the lamb out and I can see its heart is beating, but for some reason it's not breathing. These are the hardest lambs to bring round."

Farmers use various techniques to stimulate that vital first breath, including holding the newborn by its back legs and swinging it to clear the build-up of fluid in its system, firmly rubbing it clean with straw like a vigorous massage to get the lungs working and the oxygen flowing, then wiping its face and nose clean of birthing fluids to clear its airways. To an outsider, it can look quite aggressive when the farmer turns animal paramedic, but it's essential to get oxygen circulating as quickly as possible in order for the lamb to survive.

Ask any farmer and they will have their own preferred way of resuscitating a lamb if it's struggling to survive. "In addition to all the usual things, I also try moving the front legs in a circular motion to get their little shoulders moving," says David, "and gently jolting them on the ground – doing whatever I can to stimulate the lamb's first breath. You try various different things and sense what needs to be done. I'm also always picking up tips from other farmers and vets. Once when one of our cows had delivered a tiny calf and we had all but given up hope that it would survive, the vet put a few droplets of cold water in the calf's ear and it miraculously sprung to life, shaking the water out of its ears and waking up with a start. It was just incredible!"

It's awe-inspiring to see the farmers at work, calmly delivering one lamb after another, helping each newborn take its first breath, then gently placing the lamb at its mother's head so she can start licking it clean. To keep the little ones safe from infection, the farmers spray each lamb's navel with iodine as it's an open wound, and if a lamb hasn't instinctively started to suckle, the farmer will encourage it on to its mum's teat. Getting a nice full tummy of colostrum gives a lamb the best possible start in life as the liquid is high in nutrients and packed with antibodies. Whether a lamb, a kid, a calf or any other newborn, it's like the elixir of life – fast-tracking it to a healthier future.

"Colostrum's the very first milk that a mother produces," Robert explains. "And, although all of their mum's milk is important, this first milk is much more highly concentrated. It's high in protein and easily digestible, so a little goes a long way. On the farm, we've seen time after time how important it is to get colostrum into a newborn immediately after it's born. Without it, we'll have a real fight on our hands to keep a newborn alive. And colostrum is not something that can be manufactured artificially as it's tailored to each baby's specific needs."

When it comes to assisting multiple births, it's obviously less straightforward than delivering a single lamb – sometimes double or triple the potential problems. For one thing, there are more little lambs' limbs to contend with inside the ewe. For the farmers trying to assess what the problem is, it's quite literally a case of getting a feel for what's going on. "One time when I was assisting with twins, I remember thinking, 'This is a jumble in here,' because you're working blind," says David. "I was stumped as to what to do but eventually I moved the lamb at the back, which would have been born second, so it came out first, and both lambs were absolutely fine. Sometimes it's just a matter of trying something new.

"Another time, I was lambing by myself in the middle of the night and I really could have done with someone's help," he recalls. "I had lambing ropes around the newborn and I needed both of my hands to make sure the head was coming through the birth canal correctly. I didn't have time to fetch anyone else to help me, so I looped the lambing rope around my shoe and, by carefully easing my foot backwards, hey presto, the lamb came out easily."

In the dead of night, David's wife Anita will often come along to help with lambing. At first it was just to watch and keep him company, but soon she was getting involved.

"My first lambing was during the time we were courting," Anita says. "I'd watched David lambing so many times and I said, 'Can I have a go?' And it was brilliant! It's such an amazing thing helping to bring a little defenceless creature into the world. Now I help out a lot. Mind you, I know when to say 'I can't do this' and get David to help me, like when there are legs all over the place or two lambs coming out at the same time! I used to get quite emotional if a lamb didn't come round, and when they are really tiny and cold I always want to take them home with us – but David won't let me!"

"We like to leave the lambs to bond with their mums if at all possible," says David. "If they're tiny we'll put them under a heat lamp and dry them off with towels. If you can make sure their tummy is filled with milk and get them under a heat lamp, it's amazing how you can improve what looks like a lost cause. One springtime, I came up with the idea of re-using a fizzy pop bottle to make a mini hot water bottle. It saved the lamb's life."

Plus it was a good excuse for David to justify his addiction to a certain orange-labelled energy drink . . .

Because David is naturally more of a night owl and Robert likes to be up with the lark, Robert will be the first to check every area of the lambing pens in the morning. "It's very important that whoever is on shift goes round all the lambing areas as there will be some ewes who have given birth since David or I was with them," says Robert. "In each case, you do the same thing – spray the new lambs' navels and put them in a small pen with their mums to encourage the bond."

The sheep will have been marked up as to how many lambs it will be having, but occasionally it can be a bit of a guessing game as to which lamb belongs to which ewe. "Sometimes the ewe may have been mated by two different rams so its lambs could have totally different characteristics," Robert explains. "You just have to process it all in your mind – which is sometimes not that easy first thing in the morning! But the crucial factor is that the ewe is happy with the lamb, so if it's happy, it's the right lamb! The important thing is to check that the ewes are comfortable, that they have everything they need, and if they are having any difficulties, we can intervene."

With ewes only having two teats, if one has triplets, Robert and David will sometimes use a method called "wet adoption" to help bond the third triplet lamb on to another ewe that has only given birth to a single. It's not essential to have two lambs to each ewe, but it helps make sure each lamb gets enough to eat. Sometimes a ewe will reject a lamb because of the pain associated with its birth, so, when they can, farmers will always try and adopt it on.

Wet adoption is a way of catching and collecting the new ewe's birthing fluids in a basin and rubbing the fluids on to the triplet lamb. The ewe will then smell the lamb, think it's her own and bond with it accordingly. As for its other lamb, it now has a sibling! "A lamb's best interests are served by being raised by a mother or a surrogate mother," says Robert, "and adoption is very much a two-way process. We need the ewe to love the lamb, but it really helps if the lamb likes the ewe too. Farmer Kate is probably the most dedicated when it comes to trying to adopt animals on, but that's because she's the one that otherwise has to hand-feed them – that's maybe why she's so diligent!"

During the height of February lambing, around 20 ewes will give birth each day, so for the farmers it's a matter of constantly monitoring what's going on, both with the ewes that are lambing and the ones with newborns. It's a 24-hour job and, although the sheep will have been scanned, ewes don't have a definitive due date. When ewes are being mated, the farmers will fit a raddle on to the ram, which transfers a dye to the ewe indicating the day that coupling happened, but pregnancy periods vary. Like buses, nothing will come along for hours, then there may be a sudden rush. Several ewes often give birth at once.

"Once lambing starts it does seem to encourage more of the ewes to go into labour," says Robert. "When one of them starts to give birth, it definitely stimulates the maternal instinct among them and there's certainly a broodiness about them. Also, sometimes a ewe may try and steal a newborn lamb from another ewe because her hormones are in overdrive. For me, I celebrate every lamb that's born in daylight hours because it's one less that will be born at night! When I was a little boy, I'd go to bed and hope that there would be a new lamb overnight – now I just hope they'll wait while tomorrow."

In the first episode of the 2021 series of *Springtime on the Farm*, we saw Robert and David arriving to work in the Roundhouse at 3am. There were newborns everywhere. "This is the reason we don't leave them very long," Robert explained. "If you had a full night's sleep you'd come to chaos in the morning, so you just have to sleep when you can at lambing time – catch an hour here and there. It's fast-paced work but it really gets your blood pumping. There's lots happening, you have to keep your wits about you and react to whatever's thrown your way, because there's always lots of curveballs."

Robert and David check on the sheep every couple of hours during the night and every area of the farm is fitted with CCTV cameras so they always know what's happening. In the case of any emergencies, they can be at the ewe's side in minutes. They live next door to each other, just five minutes' walk away from the Roundhouse.

"When the weather is nice, lambing at night is a joy, but when it's terrible, it's a chuffing nightmare," says David. "One night I had to get from my house to the Roundhouse and there was torrential rain lashing down, the loudest thunder you can imagine and bolts of lightning going off all over the place. I felt like I was in a disaster movie. When you start your shift wet through, and you know you've got to get through the night, it can be a real challenge. On top of that, you've already done a day's work. Somehow, though, we always manage to get there in the end."

"No matter how busy it gets, there has never been a time when it's all got too much," says Robert. "We are incredibly lucky here because my dad taught us everything we know and is still always on hand if we need any advice, plus we've got a great team here with Ruth, Kate, Charlotte and the other farmers. We also have vets we can call on to help and we know when we need to do that. For me, it takes a bigger person to be able to say, 'This is beyond me, I need help', rather than struggling on and eventually just making a mess of it."

One member of the Cannon Hall Farm team has earned the nickname "Little Miss Springtime" as she has become so integral to the lambing season. All of the farmers don their waterproofs and muck in with the lambing, as does farmer Kate Bodsworth, who is generally to be found in the Reptile House. Kate is the resident reptile expert and began her career at the farm doing work experience in 2011 when she was just 16. She then studied Animal Behaviour and Welfare at university before walking straight back into a full-time job at Cannon Hall Farm.

"The first time I was asked to look after a sheep I was mucking out the meerkats and Robert came over and said, 'Do you fancy bottle-feeding a couple of lambs for us?' Well, how could I refuse? They were these two little Hebridean lambs and they were the cutest things I'd ever set eyes on. Roger showed me how to mix up the milk for them and how to hold the bottle properly and off I went. He's the best teacher in the world. I learned some agricultural subjects at uni, but it's nowhere near as useful as watching someone like Roger, who has done it all his life.

"I started off just bottle-feeding all the triplets – the ones that couldn't be adopted on to another ewe." This might happen if there are lots of triplets being born at the same time and wet adoption isn't possible. "Then I started helping out with actual lambing. I think one of the reasons I got asked is because I have really, really tiny hands, and the first time I was asked to assist was when a triplet lamb wasn't in the correct position. I had to put two hands into the ewe's womb and manipulate the baby lamb so that it was in the correct diving position. My heart was racing and I was so anxious because I thought, 'If I get this wrong, they will never ask me to do anything again.' But it lived, so I suppose it must have been a success.

"In my very first year, when I was helping out, Roger and I were milking Gloria the goat. She'd only had one kid that year and, as she was one of a breed called Anglo-Nubian, she produced a lot more milk than her kid needed. It was great news because we were also able to use her milk for any goats or lambs that hadn't got enough. That particular day, I was holding both Gloria and the bucket underneath her, and Roger was milking away, but she moved suddenly and the milk went flying everywhere – including into Roger's favourite flat cap. I felt mortified and guilty and kept apologising, and I thought, 'That's it! I'll never be asked to help out again,' but he just laughed and, in true calm-Roger style, said, 'There's no point crying over spilt milk.'"

Since the spilt milk incident, Kate has gone on to perfecting her technique for milking Gloria. "She's such a lovely patient goat and, when I started, she would be standing there as if she was thinking, 'Ah bless her, she's trying'. It's harder than it looks, though! You put the teat between your thumb and the soft part of your palm and pull it down. It takes practise but I am pretty good at it now – nowhere near as fast as Roger, of course."

—

The saying goes that where there is livestock, there's dead stock, and even with an expert team of farmers, not every new-born animal makes it. "It happens quite a lot in lambing, which is obviously really sad," Kate reveals. "One time I was assisting Roger, Robert and David – all my bosses were there at the same time, so no pressure! Immediately, though, I could tell that things weren't right. The lamb was in the right position, but it almost felt too dry. After a few minutes, I asked Dave to check, so he scrubbed up, got his hand inside the ewe and said, 'Yes, you're right, it's dead.' I was so upset because you never like losing a lamb, but I know it's par for the course in farming. It's especially hard when one of the pet lambs I bottle-feed doesn't make it. You can't help but question what you could have done differently. Should I have done this, or that?"

One condition that the farmers have to be super-vigilant about is Twin Lamb Disease, a metabolic disorder that can send a ewe into crisis from a lack of nutrients. As David explains: "The ewe may look a bit off colour one day and then, all of a sudden, it has no energy to feed the lambs and its system shuts down. And it's not just twins; the more they are having, the more likely it is to happen. A ewe can die in no time, so it's a case of continually monitoring the pregnant ewes, making sure they look well and are standing strong."

Twin Lamb Disease can be cured by injecting calcium into the ewe. Happily, they recover almost instantaneously. "You can have a ewe in the jaws of death one minute but, if you just treat them in the nick of time, you can save it," says David. "Twin Lamb Disease and prolapses are our biggest headaches with lambing, so we're continually watching for those. There's nothing worse than a ewe having three good lambs inside it and it dying in a moment."

"When we were younger, and before we opened the farm, losing any animal was a major blow for us," says Robert, "especially to my dad because, back then, every eventuality could impact our livelihood. A bad lambing meant a bad year. Even now, with so many years of farming behind us, we're always taking on guidance, especially when we introduce a new breed to the farm. We try and get as much information as we can about how to care for each particular animal," he explains. "Even two kinds of sheep will have different needs and characteristics. Humans can tell you if there's something wrong with them, but we have to act as parents for our animals. They are relying on us so we can't let them down."

With all the stress of having an animal's life in your hands, the blood, sweat and all manner of sticky fluids to deal with on a daily basis, lambing certainly isn't for the faint-hearted. As Robert explains, "It's the best job in the world, but you really appreciate your sleep when you get the chance. But when that alarm goes, you have to get up, get dressed and get back into action."

What's more, February is just the warm-up for April lambing, when the other three-quarters of the Nicholsons' ewes will give birth. And just in case they don't have enough to contend with, there's a whole host of other things going on at exactly the same time . . .

7

When the first spring lambs arrive, the goats start kidding too – much to the delight of Cannon Hall Farm fans. Ever since Roger bought the farm's very first pygmy goats in 1991, they've been a permanent part of the collection.

Dandelion and Burdock were the first kids to be born at the Nicholsons' farm, and, since then, there have been hundreds more, including the gorgeous little Millie (originally called Camilla), who captured everyone's hearts in 2020. When Millie was born, it was touch-and-go whether she would survive as she was very tiny and her mum only milked on one side. It was decided that her stronger sibling, Charles, would stay with the mum, and Millie was moved to the farm's special baby unit. There she was bottle-fed by hand every four hours until she grew and started to become stronger. By chance, a viewer had her own lonely pygmy goat called Primrose and, having heard about Millie via social media, she suggested that Primrose and Millie might be good playmates for each other. It was an instant success story when the two met, and, like the best stories, the two girls have lived happily ever after together at the farm.

There are no prizes for guessing that pygmies are the small-est breed of goat, and they're little fireballs of energy as they charge around the rare breeds barn with their friends. They even have their very own adventure playground to run, leap and slide on as the Nicholsons are conscious to provide social enrichment for their animals, as well as food and shelter.

When pygmies give birth, they generally need little assistance from farmers, which is a relief because of their diminutive size. There are other breeds of goat at Cannon Hall Farm, including Saanen and Anglo-Nubian, but these are a standard-sized goat, so it's easier to get hands on – or to be accurate, hands *in* – with them if they need help delivering their kids.

"We try not to help too much with the kidding, because there just isn't room to manoeuvre them," says David. "Also, they can stop pushing if you start to give them any help, so you have to be really sure that they need and want your intervention. Like we do with ewes, we hold back and let goats do their own thing, unless it's an obvious emergency."

"In terms of giving birth, goats are very much like sheep, just a lot smaller," says Robert. "You don't interfere with what you can't help, so we just leave them plenty of time to deliver by themselves. But when we are kidding in the middle of the night and a goat needs your attention, it's definitely an 'Oh 'eck moment'," he adds. "We know that a vet will always have a much better chance of saving a kid by emergency caesarean. This happened with little Snowdrop, who we featured in *Springtime on the Farm* in 2021." Bluebell was a first-time kidder and she was struggling when she went into labour. Three hours later, she still hadn't been able to deliver by herself. By quickly getting her to the vets, the team was able to save both beautiful Bluebell and her baby.

When Robert got Bluebell and Snowdrop back to the farm he was initially worried that the new baby wasn't feeding well and it was an anxious time for all involved. It was only when he watched the footage of Snowdrop's first few hours that he realised why she wasn't feeding – she'd already had a belly full of milk at the vets. "I was in such a panic at the time and couldn't understand why she kept refusing milk. She looked healthy enough, but that first day was a mystery to me – until I saw the programme a month later!"

Farmer Kate has hand-fed countless goat kids and lambs over the years – hence her nickname Little Miss Springtime. Making sure the newborns get the nutrition they need often involves tube-feeding – inserting a tube straight down the throat and into the kid's stomach. It sounds extreme, but it's all in a day's work for an experienced farmer like Kate. "When I first started doing it I thought, 'What if I choke this poor little thing?' But now it's something I can almost do with my eyes closed. The tube is about 30cm long and you insert it into the mouth, push past the soft palate and they kind of swallow it. Once you can hear the gurgle in the stomach you know you are in the right place and then you pour the colostrum or milk down it straight into their tummies. It's such a lovely thing to be able to do and although I've tube-fed hundreds of lambs and goats it's something I'll never tire of. Sometimes when it's really busy I go from one lamb to the next and by the time they're all fed, it's time to start all over again. There are times when it's ten o'clock at night and it's minus 3°C and you feel like you have loads of loads of animals to be fed. But that's all part of it. To me it's the best job in the world."

Tube-feeding is used when the newborn is orphaned, too weak to suckle, or can't latch on to the ewe's teat. When things get really busy and when the lambs have got the idea of sucking for milk, the farmers train them to use the automatic feeding system. It's like a vending machine for animals fitted with plastic teats for instant nourishment. "The lambs can go back and forth and have a feed whenever they need it, so it's a fantastic system," says Kate. "Until it's your turn to clean it all out! But it's vital that the automatic milking machine is always one hundred per cent safe for the newborns."

"Hand-feeding is one of the best parts of the job," says Kate. "It's really nice to have that kind of relationship with them. I'll go down to the rare breeds barn and, if there's an alpaca or a goat, they'll all come running up to me and I think, 'Ah, that's nice, it's because I raised it,' but the other part of me thinks they'll do that for anyone if they think they have food . . ."

One of the recent pygmy goats that's found a special place in everyone's hearts is little Noddy, who came into the world in February 2021. When he was born his neck wasn't strong enough to support his head and so it hung down by his chest – hence his name. The little thing looked so helpless that farmer Ruth fashioned him a splint made out of a cardboard toilet roll and a sports bandage. Robert would feed him early in the morning, Kate would take over for the feeds during the day, and David and Anita would take care of him every night. To begin with, Noddy really struggled, but with all the care and attention that he was given, he was soon on the mend. After just one week, his neck started to get stronger.

"He's such a fighter, and a real escape artist," says Kate, who helped to hand-rear him. "He lives at the Roundhouse and likes to say hello to the other goats who live next door and also the family of Swiss Valais sheep that live on the other side. He's very good at making friends so he'll often escape and do his rounds. Every year we get at least one pygmy goat that we bottle-feed who grabs our attention, mainly because we have to put so much effort into helping them in their first few months, and 2021 was definitely Noddy's year!"

Goats certainly have huge personalities – no matter how tiny they may be – so there's always a surprise in store with the goat gang at Cannon Hall Farm. Take, for example, the kidding season in 2021. As usual, the Boer billy goats would have been allowed to meet up with the Boer nannies around 150 days before the (hopefully) planned arrival the following spring. But when it came to the big day, a surprise discovery was made. Instead of finding the expected Boer kids, it seems that one of the pygmy billies had taken it upon himself to get up close and personal to no fewer than five of the (considerably) bigger ladies. Nevertheless, the newborns were as adorable as all the other kids on the farm – and, besides, who doesn't like a surprise?

In spring 2021, Gracie the maiden milker was welcomed to Cannon Hall Farm. It was a lovely episode captured in *This Week on the Farm* and followed the story of a beautiful Saanen goat who needed a new home. Gracie lived in Huddersfield and her owner Rita Hurst contacted Cannon Hall Farm because Gracie's lifelong partner had died and she had loneliness issues. "All the time Gracie seemed to be searching for him, which was quite upsetting," Rita explained. "She really looked as if her heart had broken and she seemed to be giving up on life, so I called the vet, who suggested that rehoming her might be the best option. I didn't want to let her go because I care so much about her, so I knew it would have to be a very special place."

When Rita called Cannon Hall Farm, they were only too happy to help, and it was a real bonus for everyone when Gracie arrived at her new home, as Robert reveals: "It turned out that Gracie was a maiden milker. I'd never heard the term before but apparently she came into milk without ever having a kid. As we use other goats' milk a lot on the farm to feed orphan kids, lambs and crias (alpaca and llama newborns), Gracie was an incredibly welcome addition to our farm. We then had a call from a local donkey owner whose jenny [female donkey] had a foal but she wouldn't feed it. We were able to give her some of Gracie's milk until the donkey was able to feed the foal herself."

Goat's milk is better for baby animals than cow's milk as its fat particles are smaller and therefore more easily absorbed, so having a maiden milker like Gracie was a godsend. She happily joined the goat gang and seemed to be fitting in well. "In the end, though, we didn't want to put too much pressure on her," David explains. "We could see that she had health issues and, although she rallied, we didn't want her to have to make milk unnecessarily. That would have put her another step closer to goat heaven."

Anglo-Nubian goats Dorsey and Gloria are the two most famous goat wet nurses at Cannon Hall Farm and their milk has saved several animals throughout the year, as David explains. "In spring 2021 alone, they saved a baby alpaca, several baby goats and a baby donkey; without them, they wouldn't have made it. So we think of our Anglo-Nubians as the real heroes of the farm."

—

Over in the rare breeds barn, there were more unusual animal antics in early 2021 when Robert and David conducted a scientific experiment to find out which of their female alpacas was pregnant. If you'd call a spit-off a scientific experiment, that is . . .

Alpacas are pregnant for 11 to 12 months and, as their handsome lothario Zander had already fathered five beautiful crias, Robert and David were eager to know whether there were likely to be any more. Although Zander has proved to be Robert's nemesis many a time and likes to rear up to show who's boss, Robert and David put a head collar on him and prepared for a spit-off with his girlfriends Beyoncé, Whitney, Shakira and Audrey. This would indicate whether any of the females were pregnant or were not in season and therefore unwilling to mate. One by one, the females were brought towards Zander, and if they spat at him, it was a sure sign that they were expecting.

Had they sat down in front of him, it would have meant they were ready and willing to mate.

In the alpaca and llama world, a spit-off can mean an awful lot of saliva is produced, which is particularly unpleasant as alpaca spit is a mix of saliva and gastric juice. Therefore, the experiment was not without its downsides as both Robert and David were in the vomit-scented spittle firing line. And there was a lot of it being thrown about. But as rudimentary as the experiment seemed, the results were fairly conclusive. All four alpacas appeared to be pregnant as the girls had all spat away Zander's advances.

Later in February, vet Shona Searson scanned Zander's girlfriends to confirm what the spit-off had suggested. Using ultrasound equipment, if a foetus is present, it shows up as a black circle on the scan. "There is some truth in the spitting technique," Shona explains. "The females will reject the males if pregnancy hormones are present in the body. To confirm the results of the spit-off for definite, Shona carried out an official test and yes, in this case, Zander had successfully sired all four females."

Zander is also the father of Adam, the beautiful white alpaca that was born out in the fields in 2019, and Robert, who was born in 2021 and lives with Annie, the orphaned Boer goat. The story of little Adam captured everyone's hearts, but as he has two half-sisters at Cannon Hall Farm and Zander is the dominant male, he was moved on to a new farm where he would be able to be top dog. "It wouldn't have been fair to keep him," says David. "Not only would he fight with his sisters, but he would try and mate them too. As for the little cria Robert who was bottle-fed from birth, although the Cannon Hall Farm team will hate to see him go, once he reaches maturity he too will be re-homed. It sounds harsh, but, in farming, males and females are kept separate, and supervised mating takes place to avoid interbreeding."

"Robert the alpaca would make a fantastic therapy animal,"

says Robert (his human namesake). "Like Helen, our other hand-reared alpaca, Robert is incredibly friendly and tame, so he has a very worthy future ahead of him." As with Zander's other daughters, Helen, Alpaca Chino and Shannon, once they are old enough to breed, a new male companion will be found for them. He won't be a permanent fixture though, so Zander will always be the head of the herd.

—

In spring 2021, fate threw one of those curveballs into the mix that unsettled the balance of the farming routine when news of a fresh outbreak of bird flu flew through the rural community. At the time, it didn't cause much of a stir in the national news – but perhaps that was because everyone was more concerned with a certain bigger disease that had taken hold. Besides, unlike previous variants, which could be passed on to humans, the new strain of avian flu only affected birds. Nevertheless, as the Nicholsons keep hens and cockerels at the farm, it was an issue they had to address.

"Any poultry business, or even any general households that kept pet chickens, had to take steps to prevent bird flu running rife, as it's so easily transmittable from wild birds to domestic birds," says Robert. "Therefore, even though the risk was considered fairly low, we had to take the preventative measure of keeping our birds indoors. It was a much bigger issue for commercial poultry farms, but nothing more than an inconvenience for us. Nevertheless, we wanted to keep our cockerels safe."

It was a serious issue, but Robert and David decided to tackle the subject from a more light-hearted perspective. So when it was covered for the spring 2021 series of *This Week on the Farm*, it was given a Western-inspired twist . . .

Having managed to contain the majority of their cockerels

fairly quickly, three of the crew had escaped capture and needed to be herded up by Sheriff Rob and Deputy Dave. "Butch, Sundance and Billy the Kid were used to having free rein of the farm," says David, "so taking away their liberty really ruffled their feathers." After they had cornered the feathered felons, old softies Robert and David wanted to make the cockerels' incarceration more comfortable. And knowing from personal experience that lockdown can be boring and stressful, whichever two-legged animal you happen to be, the brothers created a washing line skewered with the cockerel's favourite veggies for them to graze on. They also provided edible treats hidden in a pet puzzle for them to uncover. The foodie distraction certainly seemed to do the trick and the cockerels tucked in.

The colourful characters have become firm farm favourites since their arrival at Cannon Hall Farm in 2020. Robert and David had bought a batch of Shetland chicken eggs from a farm in York, hoping they'd boost the number of hens at the farm. Unfortunately, all but one of the hatched chickens turned out to be male. "And the sole female was killed by a fox," says Robert, "so we're not having much luck with chickens at the moment, but we'll keep on trying."

Thankfully, the bird flu lockdown didn't cause the cockerels any stress, and, a couple of months later, they were back patrolling the farmyard. It had been another drama that the Nicholsons had successfully avoided turning into a crisis.

—

The brothers' determination to keep on going, no matter the challenge, is an essential quality for farmers. Experience tells them that there will always be trying circumstances when you least need or expect them . . .

On 22 February 2018, Anticyclone Hartmut, or as it came to be known, the Beast from the East, put bad weather back on the newspaper front pages. The Arctic outbreak brought freezing cold winds from Siberia and temperatures plummeted to as low as minus 30°C. It brought havoc to Europe, with the UK being hit particularly hard.

At Cannon Hall Farm, the Beast from the East hit just as the farmers were turning out their February lambs into the fields to pasture. Although the land at Cannon Hall Farm is fairly heavy and clay-based, the fields at their other farm in the Gunthwaite Valley have sandier soil so it's better for turning out the sheep earlier in the year. "When we first found out about the better soil at Mill Farm, we knew it would be a game changer for what we do," says Roger. "The fields around Cannon Hall Farm get too soggy and puddle-filled for sheep to be out as early as February. But it's generally much better at Mill Farm."

Unfortunately, though, in 2018 not everything went to plan. "David was away skiing, and Dad and I had just moved the first lambs and their mums to Mill Farm," Robert explains. Ordinarily, they would have been perfectly happy to be out there for months, but we got caught out by the weather and they got a real belting of snow. We had to take the trailer back down to them, dig them out of the snow drifts and move them all back to the Roundhouse. After that, we were watching the weather forecast like a hawk. We'd moved them back out into Mill Farm but when we could see the next dollop of snow was coming, out we went and fetched them back. Then out again, and in again. In the end we turned out those sheep and brought them back four times. It was like the 'Hokey Cokey' for them, the poor things. And each time, there and back, it would be about ten trailer loads of around fifteen sheep each time. Plus all the roads were covered in snow and we could only get the trailer so far down the track, then we had to walk the sheep into the fields."

Although the Roundhouse is usually the ideal option for lambing, as it provides both shelter and natural air conditioning, the Beast from the East proved too much, as Robert explains. "When the wind was blowing from the east, because the sides are open, the snow was coming right into the middle of the Roundhouse. So although we were inside, we still had to contend with six inches of snow on the floor. And, because we were lambing, we had to clear the snow two or three times a day so that we could keep everything dry. It really was the worst and we had to use three times more straw than we normally would. It was just miserable and really hard work."

"At the time it felt like we were constantly watching the weather forecast and trying to beat the snows. Once we just managed to beat it by an hour. Any longer and we would never have been able to get the sheep out of the fields. We were able to keep the vast majority of the sheep alive, but it was really touch and go at the time. Certainly the most challenging time we've ever had."

Nothing seems to unduly stress any of the Nicholson farmers, but the Beast from the East was one chapter that tested them to their limit – and made them very aware of the advances that have been made in weather forecasting.

"In the old days, we used to work with an agricultural contractor called Eric Ellis," says Robert. "He was connected to the farm ever since Mum and Dad first moved here in 1959. He was the type of farmer who would never think of looking at a weather forecast – glancing out of the window would be the most he would do. He'd just carry on regardless and I am sure he was a happier soul because of it. He never stressed about the weather; he was just happy being out on his tractor, getting his work done, whether he was in the middle of the Beast from the East or a heatwave."

Although sheep can survive on snowy hilltops – the

landscape would look very different if they couldn't – farmers need access to their animals when grazing is impossible so that they can move them to where food is available. And in challenging conditions where roads are impassable, having the sheep contained in the Roundhouse is a much safer option for both the farmers and their flock.

"Forecasting has improved so much over the years that weather apps have become a lifeline for us," says Robert. "While they are never going to get it totally right, there's definitely enough information to plan for a major weather event."

Roger recalls an early incident in his farming career when he also fell foul to the weather. "I was very young and I somehow got my timings all wrong and moved the sheep out into the fields really early on in the year, thinking they would have enough grass to grow on. It had been mild, there was a good covering of grass and, to be really honest, it was because I was so broke at the time and I didn't have enough money to keep them in hay. So I figured there'd be fine outdoors. The trouble was, they finished off that field quicker than nature could make the grass grow, and I had to bring them all back in again. It was one of the lessons I learned very quickly. And there were a fair few of those when I first started out."

8

No matter how busy the Nicholsons get during the lambing season, there's always time to mark special days in the calendar. January and February can be quite bleak months, especially with Christmas behind them, so it's wonderful to have an excuse to do something different. Take Valentine's Day, for example. Although they may protest that they're not a family of romantics, Cynthia remembers the time when Roger brought her a bunch of roses for Valentine's Day and kept them in the hen house until the following morning. "The trouble was it was so cold, they froze in the bucket!" she recalls. "I've still got the petals from them though."

But there were smiles all round when Shrove Tuesday rolled around every year. "Mum was always fantastic at making pancakes," says David. "So even though we weren't in any way a particularly religious family, we'd always keep up the tradition. As for how many pancakes we'd each eat, that would depend on how many Mum was prepared to make! Dad would always get the first one, of course, then we'd line up for the next one, eat it really quickly and get back in the queue. Nothing fancy on them, mind, just lots of sugar and a squeeze of orange."

"They were like greedy little gannets, the three of them

queuing up by the side of the hob with their plates held high," says Cynthia. "I'd make a huge batch – loads of eggs straight from the hens and full fat milk from the dairy, plus flour, of course, and a pinch of salt. Then I'd whisk it up and let it stand for a few hours in the fridge – or until I couldn't stand them nagging at me any longer to make them! Then it would be a ladle of the batter for each pancake into the very, very hot lard. I'd even toss them back then – I was quite good at it, and the boys always said it helped the flavour. Roger would have the first one (I don't know why – the first one is never very good) and they'd each be back for another as soon as they'd wolfed each one down. Even now, Roger can still eat eight in one sitting. I might allow myself one if there's any batter left!"

In an episode of *This Week on the Farm*, Robert and David once went head-to-head with a pancake-making challenge. The results weren't entirely successful. And although the duelling duo might like to think they are a bit handy in the cooking department, it's eldest brother Richard who can show everyone a thing or two. Every Wednesday evening, he presents an online cookery show live from his kitchen in South Yorkshire, whipping up everything from a simple supper to a celebration feast.

"I've always had an interest in cooking," says Richard. "I suppose I was eleven or twelve when I made a Swiss roll at school and brought it back home, quite proud of my efforts. Dad was never backwards at coming forwards when it came to tasting cake so I had a willing food tester at my disposal. On another occasion I made scones and they tasted pretty good too. But I would never have guessed it would be over forty years before I baked my next batch, practising for the 5 on the Farm Festival in 2021, where I produced some lovely scones to my mum's recipe. You don't bake much when you have an expert like Mum around.

"On another occasion I remember asking my mum how to

86

make Bolognese. I still remember how she rustled up some mince, a rather ancient clove of garlic, bay leaf, tinned tomatoes and tomato purée, and patiently put them together to make a ragù. After that, I became more interested in recipe books. I now have hundreds and doing my weekly Facebook Lives has led me back to them. I love old recipe books. I have a copy of *Mrs Beeton's Book of Household Management*, bound in red and gold, which I inherited from my grandad's sister, Auntie Queenie, and it's fascinating. My mum still uses her recipe for plum jam, which we sold for a while when the first gift shop opened. An old chap and his wife came along and seemed to buy most of it. He became known as The Jam Man, an unlikely superhero, and one of many regulars who adopted us over the years.

"When it came to TV cookery programmes I loved Keith Floyd. His love of food from all around the world, his ability to get on with all types of people. His reckless abandon, the sparkle in his eye, the fact he cooked with passion, joie de vivre and a glass of good wine close to hand. He was one of the first celebrity chefs. A couple of years back, we held a day to raise money for our chef Tim Bilton, who is fighting cancer, and some top chefs turned up to offer their support. Among them was Jean-Christophe Novelli. His driver announced that he used to drive Keith Floyd around and he had a tale or two to tell. Nowadays, I'm a fan of James Martin, a Yorkshireman – with a passion for butter."

—

For all the fun of special days, when the Nicholsons like to push the boat out, there's one February anniversary that all farmers may prefer to forget – the year when foot-and-mouth disease broke out in the UK. In February 2001, cases of the deadly condition were detected in various isolated areas of England. By the following month, the disease was rife.

Anyone who recalls the headlines at the time will remember the reports of devastation across the countryside, as farmers were forced to cull and burn infected herds so that it would stop the spread. There was no treatment nor cure, and cattle, sheep and pigs all over the country caught the disease. Images of usually very stoic farmers showed a different story as many livelihoods were destroyed by the disease. Farmers had to kill hundreds of animals that they had spent their lives looking after.

Foot-and-mouth disease is highly contagious and, unlike many ailments that are spread from close contact with the infected animal, it can be passed on via contaminated clothes, animal food, farming equipment and even vehicles. At its most virulent, even driving on land where cattle are infected can pass the disease on.

At the time of the outbreak, Robert was on a family holiday in Disneyland. "We'd been planning the holiday for ages and Mum and Dad were with us," he says; "then we got a call saying foot-and-mouth disease had broken out and the farm had to immediately close to the public. When we came home, everything was eerily quiet. It was a countryside pandemic – so everything was in a version of lockdown. When the coronavirus lockdown began in 2020, it really reminded me of the foot-and-mouth episode."

Thankfully for the Nicholsons, they had zero cases of the disease at Cannon Hall Farm. "There was no foot-and-mouth at all in our area of Barnsley," says Roger. "We still had to take precautionary measures, though, so we closed the farm to the public and ensured that anything coming on to our land was disinfected. We'd discourage any unnecessary vehicles coming on to the farm, and those that had to make deliveries were cleaned and sprayed. Some farmers put big lengths of carpet down and doused them in disinfectant for visiting vehicles to drive on. We also ensured we always had clean wellies and clothing so that there was no chance we could bring infection in. In the way that sheep dip works, we dipped ourselves. I had it drilled into me at a young age how important it is to have a hygienic farm, and we put in extra measures to be doubly safe and tried to be pragmatic about the situation. We did all we could and if we had needed to do more, then we would have done. There was no question of not abiding by the rules."

Through luck, hard work and diligence, Cannon Hall Farm remained disease-free. Nonetheless, it was a worrying time for everyone – not only for the farmers who care for the animals, but for the staff who worked at the open farm visitor attraction. Would they ever be able to return to normal? There had been previous outbreaks of foot-and-mouth disease before the Nicholsons opened their farm to the public, but the 2001 epidemic was much more serious. "The whole of the countryside was closed down and it was terrible, but people did a magnificent job in the way that they never questioned what they needed to do."

It was such a successful joint effort to contain foot-and-mouth that by late spring 2001, it was announced the disease was under control and farms were able to reopen and operate as normal. The countryside breathed a huge collective sigh of relief.

The Nicholsons' Disneyland trip that coincided with the outbreak had been one of those very rare occasions when both Robert's young family and Roger and Cynthia had managed to all go away on holiday together. Nothing like the old days, of course. When the three brothers were younger, a family holiday would have meant a couple of days on the North Yorkshire coast with either Roger, or their mum Cynthia and best friend Rosemary. Back then, before the family opened the farm to the public, someone would always have to stay home to look after the animals – and also look in on the boys' grandmother Rene, who lived at the farm too. Holidays would revolve around minigolf sessions, playing in the amusement arcades, visiting zoos and fishing in rock pools. A simple life. Roger would say to them that they could either stay in a posher hotel or spend the money on amusements and sweets. They'd always go for the second option.

"Even now, going away together in spring isn't something we can do," says Robert. "David likes to go skiing in early January and I'll try and sneak in a break between the two lambing periods, but it's not always possible. We always plan our holidays around matings and deliveries so that we know that our holidays won't be detrimental to the animals. That's just the way it is and it's not a big deal for any of us. Thinking about it, I don't think people used to be as concerned with going off on holiday as they are now. In a way, it felt that there used to be a much slower pace to life – your life was your holiday because you got to decide how hard you wanted to work.

"I'm sure back then that people extracted more enjoyment from their work than they are able to now, as everyone seems to be under so much more pressure to just earn, earn, earn. I get the impression that life used to be more of a nice meander rather than the constant hectic rush it seems to be now. Some of the tasks back then meant tougher graft and now that we have people working for us on the farm, it means that we can afford to be more flexible about sharing out the workload, so it is a very different working environment, but, as my dad has always said, 'If you love your job, you'll never do a day's work in your life'."

—

In late February, while some sheep are being turned out into the fields with their new lambs, others will be being moved to the Roundhouse in readiness for the April lambing season – in the same way that the February lambers were in January. The ewes that will be lambing in Easter are scanned to assess whether they are carrying single or multiple lambs and once this is established, their feed will be supplemented as necessary. They'll then be vaccinated with Heptavac, the catch-all prevention drug for the seven main ovine diseases. This scanning, feed-checking and vaccinating stage is all part of the continuing life cycle of farming. Year in, year out, a similar pattern of events will follow.

To keep all of their farm animals healthy, the Nicholsons have to stay one step ahead of any outbreaks of potentially spreadable illnesses and diseases, and recognise the signs when something is amiss. For Roger, this means daily walks around all of the fields, both at Cannon Hall and Mill Farm – which is how he got his nickname Ranger Roger – all the while checking that every animal is behaving as it should be and isn't demonstrating any signs of stress or sickness.

Whenever Roger himself is craving a bit of peace and quiet, Mill Farm's the place he'll be. "It's my little piece of paradise," he says. "In the spring, it's so peaceful; there's a different kind of silence and the air is so clean you wish you could bottle it. You'd make a fortune. You can't think of it as work when you are walking through here and even when it's freezing cold it's still lovely. It's never not nice for me. I love Cannon Hall Farm, of course, but I think the land at Mill Farm is wonderful because it's just a bit different."

Being higher up than Cannon Hall Farm, there's more bird-life at Mill Farm, and on a clear day, the views are wonderful, stretching across and beyond Emley Moor, which has the tallest freestanding mast in the UK. At 330 metres, it's taller than The Shard in London and the Eiffel Tower. There are huge skies, vast fields and few people, and there's plenty of space to breathe, think and enjoy the quiet.

Weather permitting, by late February every year, the first of the year's new lambs will be enjoying being out in the field, so Roger will be out walking the land – he has no trouble reaching his daily 20,000 steps quota. He checks every area where a sheep might be, taking in a wide arc of farmland, across the open fields at Mill Farm, up towards the ash trees that line the stream, past the mill pond, over the stile and up towards Buttercup Hill.

"We keep a mix of different breeds at Mill Farm, including Herdwicks, Texel Cross, Zwartbles and Jacobs, and they are all happy to live together, but from time to time they seem to find each other and get back into their groups of breeds. You also find that if a lamb has been born in a certain part of the field, it will keep returning to exactly that particular place, even if it's brought back to the farm for any reason. It's something I've seen sheep do year after year so it must be instinctive to the animal.

"If I walk towards the sheep," Roger continues, "and they get up and run away from me, then they're obviously healthy. You wouldn't expect a non-domesticated animal to happily come up to you, so if an animal is a bit listless, lame or has a bit of foot rot or scouring [diarrhoea], then I know that it's a symptom of a bigger problem that needs addressing. I'll call one of the team to help and we'll get the animal back to the farm, check it over and, if it's something we can't deal with ourselves, we'll contact the vet.

"If there's a problem with a lamb, you have to act on it immediately as it can very quickly be too late – especially as it may have been suffering for a while before anyone sees it. For example, if a lamb is constantly bothering its mother, it's a sign that she might have mastitis or an infection called orf and isn't able to feed the lamb properly. Normally a lamb will go to its mum every two or three hours to feed, then happily go off and play with its friends. So if it's paying the ewe more attention than usual, or if the ewe seems to be limping, that can be a sign of mastitis. We can move both the lamb and the ewe to the safety of one of the barns and start trying to nip any sickness in the bud. A stitch in time saves nine."

"Coccidiosis is one of the biggest diseases we have to watch out for," Robert explains. "You can have a brilliant lambing time and all of the lambs are growing nicely, then all of a sudden you can get an outbreak of coccidiosis and you can lose six, eight or ten lambs really quickly. Or you have a beautiful field of lambs and then all of a sudden your flock is decimated. Orf is well named, actually, because it's awful for the lambs and the ewes. It causes scabs around the lamb's mouth and the ewe's teats and it spreads through the flock like wildfire. And if a ewe loses feeding capacity in her teats, then that's her done as a breeding animal, which is a disaster. It's awful for the animal.

"We've found that spreading the sheep out in the fields really helps and we've got better at dealing with health matters," Robert says. "A lot of that is down to Dad's daily checks. He has the ability to know instantly if something isn't quite right. He'll go into a field expecting to find something wrong and I think that's where his slightly pessimistic attitude works really well. If you look at things with a glass half-full attitude you are never going to find anything wrong. He looks at things with a glass half-empty, always assuming that there will be negatives he needs to deal with."

The term for a person in charge of livestock is a stockman, and Roger is 100 per cent dedicated to his role. "Good stockmanship isn't something that can be taught," says Robert. "It takes a lot of time and experience and Dad is the best stockman I know. He can just walk around and tell so much, just by looking at the sheep's demeanour. It's years of experience that's made him the stockman he is, as he didn't really have anyone to learn from the way that David and I learned from him. We've been lucky enough to have continual mentoring from Dad. And while he might not think that we always listen to him, it's pretty much going in. I always think, 'What would my dad do?' And the great thing is I can always ask him. Sometimes it's just a quick chat – should we move the sheep into this field, or that one? But it's always reassuring to have two people making a decision instead of just one."

Sometimes farmer Kate will accompany Roger on his daily checks around Mill Farm and gets involved with some of the less appealing spring tasks. "The first time I had to deal with fly strike it was a bit of a challenge." Fly strike is the term for a maggot infestation, which is caused by flies laying their eggs in faeces on the sheep's backside. Left untreated, the maggots can eat into the flesh of the sheep, causing disease which can kill them. "Roger immediately spotted that one of the sheep had it, and we both got to work with the shears and disinfectant spray, trying to get to the root of the problem. I just had to keep reminding myself that however bad it looks or smells, it's so much worse for the animal, so you just get on with it."

—

The bowl-shaped valley at Mill Farm ensures the sheep stay warm and sheltered as they feast on the spring grass. "In the early part of spring, we spread the sheep out over the land, and as the grass grows and the sheep get older and less susceptible to disease, you can then condense them into a smaller area," Robert explains. When the grass is growing into a thick, lush pasture, it needs to be kept short, because once it goes to seed it doesn't produce as well. "If it's about four inches long, it will always be nice and leafy. Once the grass is growing faster than the sheep can eat it, we merge two flocks together and close one of the fields off so it can develop into hay. This will later be harvested and fed to the animals. Every year your timing is different for when you take the sheep out and then merge them because of the weather conditions, so you can't say exactly when it's going to happen. You just have to use your instincts."

When Roger was younger and the family had less livestock, their fields would be used to grow crops – barley, wheat, potatoes, turnips and oats – which would be rotated so that any blight didn't have a chance to build up. Moving crops around is good for the land and results in healthier produce. "Animals are generally kept on your grassland," says Roger, "but, to change things up a bit, you might grow a crop of mustard and rape and feed your sheep on that. They'll eat it, and their manure will replenish the earth, so that the field can then be used to grow a different cereal crop." It's important to rest the fields so that they get a break from being used to nurture sheep. "Ideally we'll turn the sheep out into a field knowing that it won't have a worm burden," says Roger. "Any worm eggs that will have been in sheep faeces will have died away, so they won't infect the sheep. If you have done your job right you know that those sheep have a good chance of doing really nicely."

In the seventeenth century, Charles Townshend – or 2nd Viscount Townshend to give him his full title – introduced a farming system to the UK that rotated crops on a four-year basis and used turnips and clover as two of the crops in his rotation. It became so widely used that the viscount got the nickname Turnip Townshend. His method for rotating crops still stands true to this day.

"We have a three-year rotation," says Robert, "so one year in three we will rotate between grass and turnips, two fields of grass to one of turnips. Some of the fields we keep as permanent pasture as we are trying to capture all of the possible carbon we can, and use our land in the most effective way for our business."

With 32 fields at their disposal at Cannon Hall Farm, the Nicholsons can decide which fields will be better for grazing at different times of the year – upland areas being dryer and therefore of use earlier in the spring than lower fields. Roger still has the 1947 map that not only shows the location of all the fields at Cannon Hall Farm, but also their original names, several of which are still used today, such as the White Mare Field and the Park fields. Many areas are called Shutts, which historically was the Yorkshire name for a division of land or an enclosure of a farm field.

When the Nicholson family bought the property and land at Mill Farm in 2004, they added an extra 65 acres to their existing farm, enabling them to vary the timings in which they farm, due to the geology of the land. Now, depending on the weather – as so much of farming is – sometimes barley can be sown in February as it needs four and a half months to harvest from seed. On an arable farm, planting barley – or drilling as it's known – is considered to be the most important job in the whole of spring.

To drill the barley, the box mechanism that distributes the seed would be pre-set so that the correct amount of seed would flow down the chute and into the ground as the tractor went along. It could be set to release from 8 to 14 stone of seed per acre, depending on how much of the crop your land would yield. In Roger's case, it would be about 12 stone per acre and 30 acres would be drilled.

"Before we were married, I'd ride on the back of the drill," says Cynthia. "It was my job to pull the handle on the side of the drill when we got to the end of each length of the field. Then the tractor would turn and set off again. I'd be concentrating on pulling the lever at the right time, otherwise they'd be a bald patch in the land come summer. And Roger wouldn't like that!"

"In January and February you seem to have all this time when it's wet and horrible, then suddenly one day it appears to dry up miraculously," says Roger. "That's when you can think about working up the land. I've had years when I could drill barley in February, but other years can be much later. In 1970, the year that David was born, I was drilling on the twenty-second of April."

This memory is still a bit of a sore point as far as Cynthia is concerned: "There was I was at home giving birth, and Roger buggered off to do the drilling!" She is still smarting from the episode. "He claims he thought it was better to leave the birthing to the experts, but some may say that's a cowardly approach. Actually, he also managed to be in hospital when I was giving birth to Robert, so he wasn't around for that birth either!"

"I had an ingrowing toenail and it was a medical emergency!" says Roger. "It was very painful. You were only having a child." Then he quickly adds, "I didn't mean that really!"

There's always a fair amount of banter shared between the Nicholson family. Fans of their television shows will know that Robert and David like nothing more than a verbal sparring session, and Richard will happily join in the ribbing when the three brothers get together. As for Cynthia and Roger, the two of them make a great double act, but that's what over 50 years of marriage gets you. After all, it was Cynthia's dry sense of humour that attracted young Roger all those years ago.

9

February is the anniversary of the arrival of one of the most famous animals at Cannon Hall Farm. Back in 2019, Roger travelled up to Oban in Scotland with Yorkshire Vet Peter Wright to bid for a Highland cow at auction. The two elder statesmen were TV gold as *Springtime on the Farm* viewers watched the two of them in action. As they headed to the market, Roger said, "The idea is to be very non-committal, even if you're very keen on an animal." Peter quipped: "Just go in there as if you have very short arms and long pockets!"

The two of them carefully looked around the Highland cows that were for sale, checking out each one's physique, gait, and all the tick list attributes needed for a top-quality Highland heifer. Well-curved horns, good teats, square back, thick coat and a well-behaved temperament. To a layperson, all of the Highlands that were up for auction looked like fine specimens of the breed, but it was love at first sight when Roger and Peter clapped eyes on four-year-old Highland heifer Fern. In fact, she made such an impression on Roger that when it came to bidding for Fern, he made sure to seal the deal with an extra £100 over the highest bid. Once Roger knows what he wants, there's no stopping him. Since then, the story of Fern and her fine son Ted has continued to delight Cannon Hall Farm fans.

Barely six months after buying the Highland duo, the family decided that Fern was such a fine example of the breed that she should be shown off at the Great Yorkshire Show. Ted would also be at her side to show what a great mum she is. The Great Yorkshire Show is known as the Oscars of farming and takes place every summer in Harrogate. The Nicholsons have a family legacy of success at country shows and this time it would be David who'd don the white handler's coat on behalf of Cannon Hall Farm.

So just a few months after Fern and Ted's arrival, they were getting ready for their debut at the 2019 Great Yorkshire Show. It took a bit of time to get Fern ready for her time in the spotlight. Luckily, David's wife Anita is a hairdresser, so she could help get Fern's coat looking like she'd just stepped out of the salon. The team also needed to get the beautified cow used to being led around so that she would perform her best on the day. In addition, Ted got a mini makeover – though not much was needed as he was such a handsome chap already.

On the big day, David and farmer Ruth walked the dynamic duo out in front of the judges and it wasn't long until Fern was being awarded first prize. "It doesn't get better than that," said David, grinning from ear to ear.

The Great Yorkshire Show was cancelled in 2020 because of Covid-19, but it was back with a vengeance in 2021. This time it was Ted's chance for glory. At two years old, and having never been shown on his own before, the family didn't have great expectations for his success. And when both David and farmer Ruth had to pull out of the show because of injury and a replacement handler had to be found, they almost decided not to compete at all. Luckily, show handler Emma Haley came to the rescue, and the family all went along to watch Team Ted from the sidelines.

Things didn't look good. Ted wasn't happy in the spotlight and there were a few "oh 'ecks" from Robert as Ted tugged this way and that, and the young bull bellowed his displeasure. Poor Emma looked exhausted when the ordeal was over and she was able to steer Ted back to the entrants tent. But, lo and behold, Cannon Hall Farm did it again. Ted was awarded first place in his class and the family were delighted – if a little surprised by the result. "We were all flabbergasted, to be honest," says Roger. "At one stage I thought he would be disqualified because he was leading the handler a right merry dance. But obviously the judges saw something special about Ted and he did look fantastic, even if he was a naughty blighter on the day."

A short while later, Ted tackled another milestone in his life when he became a fully fledged bull for the first time. "He was born in spring 2019 and we decided the time was right for him to mate with our Highland Emma," says Robert. "She's one of our other prize-winning cows and she is older than Ted but she was Ted's first conquest. She came into bulling [fertility] so she was ready for mating, but poor Ted got a bit confused and started jumping on the wrong end. He was getting all hot and bothered and I think he was in danger of giving up. Plus I was a bit concerned for him; with horns like Emma's, he could have done himself an injury, so Dave and I got Emma into a halter and held her steady, turned her around the right way so that Ted could jump on." Thankfully, it all went to plan after that. "I kept watching him and he served her three times," says Robert. "I know I must have looked like a right weirdo watching him, but I wanted to make sure it had worked and that he wasn't going to injure himself. We'll have to wait for Emma to be scanned to see if it worked and if she's in calf, but it all went in the right direction – eventually."

—

In other four-legged celebrations, February is also the anniversary of a dramatic episode in the life of Prince, one of the two original reindeer at Cannon Hall Farm. Like Fern and Ted, Prince and Jeffrey had come to live at the farm in 2019 and seemed to be settling in well. But the following February the farmers had to call in the vet as Prince had an alarming swelling on his right eye. It turned out that he had an ulcerated tumour and it was decided that his right eye should be completely removed in case the tumour spread to other parts of his body.

"It was a traumatic time for us – not to mention Prince," says Robert. The actual surgery wasn't life-threatening, but careful aftercare was essential. A cosy post-op shelter was provided for him, complete with a duvet and a heat lamp, and, within hours, the one-eyed reindeer seemed on his road to recovery – if a little confused by his new circumstances. "From then on, we made sure not to approach Prince from his blindside so that he wouldn't be spooked, but within days he seemed to have completely adapted to life with only one eye."

Unfortunately, that wasn't to be the end of the drama for Prince. During late spring in 2021, he lost another body part when a second tumour was discovered. "What started out as a little pimple started to grow alarmingly quickly," says Robert. "It became a big angry red mass, but luckily it was right on the tip of his ear, so it was a very straightforward surgery." Matt, the vet who had helped Peter Wright perform the surgery on Prince's eye, came along to Cannon Hall Farm to operate on Prince, and Robert assisted him by cauterising the wound. Once again, Prince had to have a general anaesthetic while he was having his ear removed and given careful post-op care with a cosy duvet to keep him warm. Thankfully, he was back on his feet again in no time at all and showed no sign of distress. He might look a bit more lopsided now, but he's still a handsome reindeer.

Even though traditionally we may think of a reindeer as a Christmas animal, February seems to be the month when they make big news at Cannon Hall Farm, and in 2021, there was definitely something to celebrate. Two new young reindeer, Dasher and Dancer, had arrived the previous winter and Dancer was believed to be expecting, but as with many animals, the actual day she was going to give birth was hazy.

"One day during lockdown in spring 2021, Mum and Dad were having a walk around the farm while it was closed," Robert explains, "They had both been shielding because of coronavirus, so Mum hadn't been able to see much of the farm for a while. During their rounds they looked in on the new reindeers and there, at Dancer's feet, was the most beautiful baby reindeer. He was pearly white and grey and the size of a big lamb. Dad came to tell me the news and we immediately moved Dancer's brother Dasher to a next-door pen so that we could get a closer look at the beautiful new baby. It was amazing how trusting Dancer was as we checked over her little boy. Many new mothers in the animal kingdom will naturally be aggressively protective of their young, but Dancer was happy enough for us to examine him. Her fawn had impossibly long legs, huge Bambi eyes and seemed to have no trouble latching on to his mum for a feed. It was the most wonderful sight and a really special occasion for us as it was our first baby reindeer born at Cannon Hall Farm."

The farmers are hoping that one day there will be another pitter-patter of tiny hooves, but this will mean taking Dancer to meet an "entire" boy. Prince, Jeffrey and Dasher are all castrated because fully male reindeer are too dangerous and can become very aggressive – not the thing to see on a family farm that's open to the public.

As so often happens with a new birth at the farm, followers on social media were asked to come up with a name for the new reindeer addition. He certainly had a shiny nose, so, even though it was a bit of a reindeer cliché, the name Rudolph could have clinched it, but as Ranger Roger had been the first to find it, that was what the fans voted for. If it had been a new female, it may have been called Cynthia . . .

"I've had a few animals named after me," says Cynthia. "There's Cynthia the donkey who still lives on the farm now and was born on my birthday, the third of July. Then there was a beef Chianina cow named after me about forty years ago. We used to keep Chianinas, which are the biggest breed in the world, and at that time Roger was breeding them, so he went through the alphabet naming each one. A bit like naming hurricanes! I can't remember the names of any of the others but I know one of them went for a wander one night – and ended up climbing through my mother-in-law's bedroom window. She got the fright of her life when she woke up with his front feet on her bed. She wasn't best pleased. My namesake, Cynthia, would never have done such a thing."

—

Depending on the weather (that old chestnut), from late February, it's time for the cows at Cannon Hall Farm to be returned to the fields for their summer holiday. All those months of being kept indoors might have kept them warm and cosy – with food and water on tap – but, like all of us, they need a change of scenery. "Traditionally we would turn the cattle out on May first," says Robert, "but when there's a reasonable chance of good weather, there's nothing nicer than seeing the cows with their calves charging out on to the fields and their feet touching turf for the first time. If you have higher fields you can certainly get them out earlier as the soil is more free-draining. This means you can also leave them out longer later in the year, which reduces the cost of bedding and makes for better cow health."

In the same way that the sheep are moved, the farmers carefully load up the cows and their calves into a trailer and transport each batch of Shorthorns to their new home. "Sometimes it's hard to convince them to leave the cow barn," says Roger. "So it might take a bit of a shoulder to get them up the ramp, but, within minutes of being in their new field, they seem much happier. When we first got the Anniversary Herd of Shorthorns, we had to keep them indoors for a while as a kind of quarantine, but they now always live in the Park Field that borders the parkland and reaches down towards the river. When their hooves touch the new spring grass every year, they gad about like crazy, rushing around and making lots of noise, but after about half an hour they settle down, eat grass and just do what cows do."

PART THREE
March

10

As February moves into March, you can sense the rhythm of the farm really begins to pick up. There are new lambs, kids, calves, piglets and crias in abundance, the number of visitor numbers continues to increase and, as the days get longer, everywhere there's a real sense that spring is in the air. You can almost taste it. Tête-à-tête daffodils and crocus bulbs that will have been planted by green-fingered Cynthia the previous autumn adorn the borders and pots around the farmhouse in a blaze of glory. Furry catkins appear on the trees and the birds get busier with nest building. It's a hive of activity for nature – and it's about to get even more exciting . . .

March is the month of the spring equinox, when the clocks go forward an hour and we have longer, lighter days. The word "equinox" means "equal night", when the sun is positioned directly above the equator, so day and night are the same length. This only occurs twice a year – around 20 March, and in September, around the 23rd. Meteorologists say that spring technically begins on 1 March and in farming the first day of spring is 1 April – as traditionally that's when spring lambing would begin. But whatever the actual date spring officially begins, the main shift in proceedings is that clocks move ahead one hour. There's one less hour on the day that the clocks change, but it's lighter for longer, which is a lift to everyone's spirits.

It's a common misconception that the idea of daylight saving was first thought up by Benjamin Franklin in 1784. "Apparently he thought if people got up earlier it would save on candles," says Robert. "Anyway, whoever came up with it might have found it useful back then, but I don't really know why we have to do it now. After all, it's not as if we don't have electric lights everywhere!"

Nevertheless, the clocks moving forward is as much a feature of spring as the blossom on the trees. And if it makes for slightly easier days, then it's a relief for everyone. Children don't have to get home from school in the dark and it's lighter for farmers when they start work in the mornings. Spring has sprung and everyone's happy. Or are they?

"I've noticed that the seasons have really changed during my lifetime," says Robert. "And the planet's been around for ages so something must be going wrong. Everything seems to happen about a month earlier than it used to when we were kids. When they were growing up, our children got virtually no snow in winter, so they never got snow days off school, which I always thought was a bit of a shame. A snow day was the highlight of the year when we were kids! The trouble is, I must have gone on about it a bit too much because one year my daughter's teacher called me to check we were all OK. Apparently Katie had said we were snowed in at the farm and she couldn't get to school that day. And we'd only had a little flurry! Certainly not enough for her to have a day off." Nice try, Katie.

As well as it being the month of daylight saving, there's a marked difference in the amount of birdsong that can be heard all around Britain in March. Male birds are staking their claim to territory and showing off to the females in readiness to mate.

From their kitchen window, Roger and Cynthia get a bird's-eye view of the action unfolding. "It's wonderful watching the spring birds doing their courtship rituals," says Roger. "They fluff up their wings as if they're doing a Spanish paso doble and the male birds chase each other off as they vie for the hens. They really put on a show. Even the little birds like sparrows and goldfinches do it, puffing themselves up and fluttering their wings."

Being in the heart of the stunning South Yorkshire countryside, the Nicholsons are blessed with an array of garden birds that visit all year round, like the dunnock that can often be found shuffling around the flower beds and shrubs. The dunnock is usually solitary, but in spring, when it's time for them to mate, it will steel itself, and if a rival male comes along, there's all sorts of flicking and clicking noises between them as they go head-to-head for the female's attention.

"In spring 2020, it became such a joy for us to see all the birds on the farm," says Roger. "It was such a worrying time that being able to focus on the beauty of the wildlife around us instead of fretting about Covid-19 was a great distraction. You could tune out of the bad news for a few minutes and tune into nature."

With the return of migrating birds and garden favourites practising their courtship routines, there are more voices to add to the dawn chorus choir. The feathered alarm clocks don't wait for the clocks to go forward before they announce their presence. Skylarks, song thrushes, robins and blackbirds are the earliest risers while wrens and warblers like a bit of a lie-in and won't provide their backing vocals until later in the morning. The volume hits its crescendo at sunrise before quietening down a little. Then the less welcome electronic alarm clocks take over the morning.

—

When life was put on hold by Covid-19, watching spring unfold in all of its glory became more important than ever, as it was sometimes difficult to see hope on the horizon. Rumours of the killer virus became headline news, with vulnerable people being told to shield at home and life-changing completely for many people. But it wasn't until 23 March 2020 that Prime Minister Boris Johnson announced there was to be a national lockdown. He said that everyone – not just the vulnerable – should stay at home and should only leave their houses for limited reasons. Many businesses had to close immediately "Until Further Notice". For a business like Cannon Hall Farm, whose viability relies on the visiting public, it was very grave news indeed.

"We were all terrified at the time," says Robert. "Terrified that the frailer members of the family would contract the virus, but also in fear for what could happen if the situation continued long-term. We had a certain amount of money in the bank, so it wasn't an immediate financial worry for us, and we knew we could continue to pay staff and all our contractors for a while, but we were conscious that our money wouldn't last forever. We had a big team of people whose livelihoods depended on Cannon Hall Farm."

With a number of staff on both the farming side and in retail, catering and admin, depending on how long the situation was to continue, there were a lot of jobs hanging in the balance. At the time, the general directive from the government was that people should work from home if at all possible. For some people this wasn't a problem – swapping your office desk for the kitchen table wasn't a big deal, thanks to email and shared networks, but it wasn't exactly ideal if you made your living from waiting on tables. And why would a car park attendant be needed when there are no visitors to assist? It was time for the Nicholsons to do some strategic thinking.

"Our farm shop became the saviour of our business," says Richard. "Because of people panicking and stockpiling, some of the bigger supermarkets were having difficulty supplying the demands of their customers, so farm shops like ours became more widely used than normal. We expanded our range of groceries and fresh produce and the demand for click-and-collect orders went through the roof. As a result, we were able to move staff from their usual roles in our cafés and the White Bull restaurant into positions in the farm shop. Then when the furlough scheme came along, the jobs of other people on the farm that were hanging in the balance became more secure."

Nevertheless, Cannon Hall Farm was a very different place. "It was like the *Mary Celeste*," says David's wife Anita. "It was so depressing as gradually all of the restaurant food went out of date and had to be thrown away." Roger and Cynthia, Robert and Julie and David and Anita live next door to each other in three houses on the farm, so they used up as much of the catering produce as they could, but eventually there was no use for much of the restaurant stock. "Things like gallons and gallons of ice cream had to be thrown away," Anita continues, "and we had stores full of other catering produce that couldn't be sold in the farm shop and had to be chucked – thousands of pounds' worth of food all had to go. It might seem very wasteful, but we were unable to just give the food away for health and safety reasons as it was all out of date. I'd be manning the office and the phone would be constantly ringing with people wanting refunds on their booked tickets. One day I saw thousands of pounds going back. Meanwhile, we still had to pay processing fees to the company that issued the tickets. It was soul-destroying because we'd all worked so hard to grow the business and make it a success and I could just see it dying in front of me. Every time there was a new announcement about the virus on the news, David and I would go for a walk and I'd have a cry. It all seemed so bleak."

"At one point it felt like it was going on for ever and a day," says Robert. "As for the illness itself, if there was ever one person who was vulnerable to a dose of Covid-19, it was my mum, and, as her health became more frail, it was vital that she and my dad shielded. Because we live next door, we were able to shout across the path to them to stay in touch. But I had to keep telling my dad off because he was getting too close to people. He wanted to do his job as normal, but normality was on hold."

Thankfully, it was life as normal for the animals that live at Cannon Hall Farm. They still needed to be fed, watered and mucked out every day, and their health and wellbeing was as important as it ever was. There was the usual quota of scheduled births and surprise arrivals that happen every spring; meanwhile, the more colourful farm favourites continued to do what comes naturally to them. Pygmy goats carried on enjoying their adventure playground, the meerkats kept on lookout duty and the alpaca posse looked as beguiling as ever, like a bunch of opinionated ladies putting the world to rights.

While in the rest of the world it was chaos, in the stunning setting of South Yorkshire, spring was coming into its own. "With spring unfolding around us, it seemed such a shame that our loyal supporters would be missing out on so much," says Robert. "At that time we had around five hundred pregnant ewes, but there would be nobody to see their lambs coming into the world. So I made the decision to do a live broadcast on Facebook every morning while we were in lockdown. I hoped that people would still feel part of the farm, even though they couldn't physically visit us in person." Every day, come rain or shine, Robert would be there on-screen, filling in everyone on life in lockdown at Cannon Hall Farm.

Everyone was talking about the "new normal" and Robert's

new normal routine would see him getting up as usual between 3 and 4am to do his farm rounds, checking that all of the animals looked healthy and happy. Then he'd make a list of the day's tasks ahead. Coming into spring, there would be ewes, cows and nannies gearing up to give birth, as well as several other animals with newborns, plus all the other breeds on the farm that might also be due to deliver and that need careful monitoring. With his rounds done, he'd reward himself with a bacon sandwich ("Grilled, not fried. I didn't want to end up like a walrus!"), a pint of Yorkshire tea and a nice hot bath. Then it was time to get dressed and ready for his daily 7.30am live social media broadcast.

"It became a daily ritual for me and, even when the first lockdown period was over, I continued to do my broadcasts," says Robert. "People have told me since then that it was the highlight of their day during the lockdown period – something that they could get invested in and something to feel good about. The thing is, it was the same for me – keeping that normality going, even though the real situation was anything but normal.

"From the comments that people would post on the screen as I was talking, I could see that people were getting something from what was essentially just my daily ramblings! That made me feel so good because I felt I was actually doing something useful for people to lift their spirits. Some days felt incredibly difficult – it was impossible not to feel worried about the virus and devastated by the terrifying number of deaths, but we had to stay positive and keep hoping for better times ahead."

For everyone, it seemed like every time you switched on the radio or the TV, there was more bad news, so there was a real appetite for anything that offered time out from the day-to-day routine. Some took to perfecting their banana bread recipe or thinking up quiz questions on Zoom, and entire catalogues of box sets got watched on Netflix. Distracting TV was very much the order of the day and what could be better than a passport to pastoral perfection at Cannon Hall Farm?

Meanwhile, Cannon Hall Farm's supporters were helping to keep the business afloat with the rise of online shopping. With the nation embracing a new relaxed home daily "uniform" of jogging bottoms, T-shirts and hoodies, sales of their fantastic merchandise were going through the roof. People loved seeing the brothers' catchphrases emblazoned on bright-coloured clothing, mugs, tote bags and aprons, so everyone mucked in to help get the orders to the post office. The team had launched animal adoptions in March 2020 as well, which had really taken off.

Also in March, filming for the new series of *Springtime on the Farm* was well underway. Some reports had already been completed in January and February, but with the new lockdown rules, there was another hurdle to consider – not being able to work with people outside your own bubble. It could have meant that, like many programmes, *Springtime on the Farm* would have to go on hold while the pandemic raged. But then the TV production company came up with a solution. With the magic of technology, the presenting team could not only link up to Cannon Hall Farm for the live shows, but as Robert, David and other team members were so adept at filming themselves, they could make reports for the show via the phone, all the while sticking to government guidelines.

When the show went live the following month, it was a resounding success. Granted, it looked a bit different with presenter Helen Skelton coming live from her back garden in Yorkshire and fellow presenter Adam Henson at his farm in the Cotswolds. But the team was determined that the show must – and would – go on.

"This would never have been possible without the advances in phone technology," says Daisybeck's Paul Stead. "Initially

Channel 5 said that, because of Covid-19, the programme had to be cancelled, but I begged the channel to give me twenty-four hours to work out how it could be done. Part of the production team had been using mobile phones to record pilots [one-off programmes] so we'd tried out the technology and knew it worked. So then it was a matter of sending out new state-of-the-art mobiles to the presenters Helen and Adam, and Rob and Dave, and showing them all how to record and upload material to the cloud.

"To make the programme, I would sit in front of a computer in my dining room watching the WhatsApp calls between the four of them, and another editor/producer was watching the same footage from his dining room at the same time. The programme was then edited together with the reports that were already pre-recorded and in the bag." Channel 5 gave it the thumbs up and it was all systems go. But would TV audiences like what they saw?

Viewing figures are calculated by the Broadcasters' Audience Research Board (or BARB), who compile an estimate of how many people are watching each programme via information from special BARB boxes installed in homes around the UK. In 2020, around 5,100 homes had BARB boxes, each representing around 5,000 viewers. Their viewing habits give an indication as to the total number of viewers watching each particular programme. Within 24 hours of a programme being broadcast, television companies are able to see the viewing figures/ratings of their shows – known as "overnights" – and, after 28 days, the number of times a programme is viewed online or via Catch Up is also added.

And when the viewing figures came through for the new remotely filmed version of *Springtime on the Farm*, it was clear audiences were obviously loving it. It was such a success, in fact, that Channel 5 wanted to see more of the gang and *This*

Week on the Farm was born, this time with Helen Skelton and Jules Hudson at the helm.

"I think the secret of the programmes' success is quite simply that people love the Nicholsons," says Paul Stead. "They are not acting, they have naturally warm personalities and they are passionate about the job they do. Their animal husbandry is superb and viewers engage with the whole family. What could be better?"

The feedback from *On the Farm* fans and social media followers has continued to be so positive that it's become an integral part of the daily life at Cannon Hall Farm. Richard Nicholson helps to head up the social media team and regularly communicates with farm fans and supporters. "One of the joys of social media is the constant discussion that goes on between our farmers and those that visit the farm. We get lots of messages, suggestions, enquiries – and occasionally even a telling-off – from our huge army of supporters. It keeps things real and is a wonderful way to communicate with the people who matter. The internet often gets a bad press, but for us it's an absolute godsend."

11

Turning back the years to when the young brothers would have turned the calendar over to a new month, in the old days March would have meant one step closer to their favourite hobby . . .

"We were all absolutely nuts about pond life when we were kids," says Richard, the family's number one amphibian and fishing fan. "One of my parents' first memories is of me as a toddler grovelling around the grates at Tower Cottage, where we used to live. I must have only been about two years old because Robert was still a tiny baby, and I'd go out there with my little yellow bucket and stick my hand down the drain. It was always full of old rotten leaves, but there'd be great crested newts, common frogs and toads, and I'd fill my bucket and have a good look at them. I wasn't the sort of child to handle them too much or torture them, or anything horrible like that, and I'd always make sure I put them back in the drain safe and well. I was just fascinated by them. Still am!"

When Robert and David grew big enough to share in Richard's fascination, their big brother would show them where to find their local amphibian friends. "I remember taking Robert and David to look for newts in parts of the drystone walling around the farm," he says. "It was probably the middle of winter and the newts and frogs were hibernating there, but Robert, David and I would carefully pick them out of the crevices and carry them back to the pond where we thought they needed to be. It probably killed them because the water would have been freezing cold, but we thought we were doing the right thing."

It was watching frogs and toads develop from eggs that really captured the brothers' imagination. To inquisitive little children, clusters of frogspawn and strings of toadspawn can disgust and delight in equal measure. "The fact that you can see the shape of an animal emerging right in front of you was just mind-blowing for us as kids," says Richard. "What starts out as this disgusting-looking goo is filled with life, and, in just a few weeks, you see the legs develop and then the rest of the body. We'd always seen baby animals fully formed at birth, but for us this was so much more exciting."

Frogs, toads and newts will normally begin to appear when the weather starts to get warmer as they know that there will be enough food around for their tadpoles to eat. If it's particularly warm, this can even happen in late February. Frogs return to the ponds where they were born, where they will spawn and lay eggs near the water's edge where it's sunniest. And, although toads prefer a pond, any water will do – a rain-filled ditch or even a decent-sized puddle will work if it's warm enough.

This is the reason why you might suddenly see armies of the plucky creatures attempting to crossroads in spring. Not far behind you might also spy the Lesser Spotted David in his high-vis Toad Patrol tabard. "We set up Toad Patrol because whenever you get particularly heavy rain, the main roads get covered in toads," David explains, "so it's vital we do as much as we can to help them survive." He'll be there stopping the traffic and carrying the warty wanderers to the safety of the side of the road.

The annual Great Toad Migration was highlighted in the spring 2021 series of *This Week on the Farm*. It's estimated that 20 tonnes of toads are killed on the roads in the UK every year, so the brothers came up with another idea to help reduce the road crossing carnage. They headed to an area on their farmland called Deffer Wood to build a toad motel, with the very fancy technical name of a Hibernaculum.

As chief amphibian fan, Richard explained the plan. "Although the toads and frogs would be heading for the water in spring and therefore wouldn't need the Hibernaculum until late autumn, we found somewhere that would catch the sun when the next spring arrived so they'd know when it was time to check out of their toad motel. Meanwhile, the hide would be nice and warm for them when they needed to shelter in it."

After digging a shallow square-shaped hole around a metre wide, the brothers filled it with layers of discarded pieces of pipe, sticks and hollow tubes, then topped it with chunky logs and covered it with layers of leaves to provide insulation. Although Robert protested that it was just a pile of sticks and not the boutique hotel that he originally had in mind, Richard explained why it would be the amphibian equivalent of The Ritz. "It's actually a carefully engineered habitat," he said. "The layers of sticks and leaves provide a layer to keep it warm, and organic creatures such as woodlice and spiders love to live in that environment. Frogs and toads will eat the insects, so it's a great little all-in-one ecosystem."

Toads are recognised as an indicator species, so when they suddenly appear in great numbers, it means that the environment is really healthy. They feast on insects and, in turn, larger animals eat the toads, so it's an indication that everything in the environment is working as it should.

Warty toads and bug-eyed frogs may not rate as highly in the looks department as baby lambs and goats, but they're just as cherished at Cannon Hall Farm. As well as amphibians, the brothers have also been life-long fans of reptiles and, in 2017, they opened a purpose-built area to house a new collection. Although it hasn't yet been featured in *Springtime* or *This Week on the Farm*, the incredible Reptile House is really popular with visitors to the farm and is home to all manner of slippery, scaly and multi-legged creatures – including a bearded dragon called Sherbet, boa constrictors Blip and Fergus, and a curious soft-shelled turtle called Piggle.

As all of the animals that live in the Reptile House originally hail from more exotic climes than Barnsley, they live in tanks that replicate their natural habitats. For this reason, the British spring doesn't factor in their individual life cycles. Depending on their age and breed, they'll happily birth all year round. But it would seem that spring fever does indeed sometimes spread to the Reptile House. "Iggy, our male iguana, definitely gets a bit more dominant in spring," says Kate (aka Little Miss Springtime). "He does a lot more displaying in spring; he'll nod his head up and down and do a lot more posing."

—

Elsewhere on the farm, other animals are gearing up for spring mating too. Pairs of buzzards can be seen over the South Yorkshire hills perfecting their spectacular aerial display. The male soars into the sky then free falls down to impress his partner in a rollercoaster courtship. Sometimes the two birds lock talons and spiral downwards together. At the farm, hedgerows' buds start opening up to attract pollinators and, as the insect world increases, there's a buzz in the bee kingdom and the volume increases on the dawn chorus as birds greet the early light.

At Cannon Hall Farm, during the two main lambing periods

of mid-February and Easter, the farmers will try and time a new arrival for March, and, in 2020, the stars aligned for the imminent arrival of a foal to their magnificent Shire horse Orchid. It was an anxious time for the farmers as the first attempt to breed a Shire horse foal had gone badly wrong . . .

Back in 2017, the Nicholsons' Shire horse Poppy gave birth to a beautiful foal they named Chester. It was a difficult birth as Poppy was overdue and had developed an edema – a retention of fluid trapped around her body. "We were really up against it," says Robert. "The vet came out every day and we repeatedly asked if we could induce the birth because we knew Poppy was on a knife edge, but inducing labour in horses often doesn't end well. So the fact we got her through to the conclusion of her pregnancy was a big achievement in itself." Unfortunately, though, the beautiful mare was unable to get back on her feet again and, despite the team's best efforts – and even with the intervention of the fire brigade – she didn't make it.

Meanwhile, Chester the foal was doing tremendously well. "We gave him all of the colostrum that we could from Poppy and he was drinking plenty of milk and he looked hale and hearty, but then he showed signs that something wasn't quite right," Robert explains. "We immediately took him to the top equine hospital in Yorkshire, which is more accustomed to treating world-class thoroughbreds than Shire horses. He was fed well and given blood infusions and each time he seemed to rally, and each time we thought maybe we'd got half a chance in saving him. But then, he suddenly developed septicaemia and died. By the time you realise what's happening it's usually too late to do anything about it because septicaemia is a silent killer, which just crept up on him like a fog. We did our very best for him, but sometimes your best just isn't good enough."

When Robert reported on the events on his Sunday social media broadcast, a record number of followers were watching.

"It really was harrowing," he says. "I was looking at how many people were online at the time – and I could see the numbers increasing – and I'm there holding back the tears and whimpering the news. I was devastated that all our efforts were in vain and that we'd lost both the mare and the foal – and I felt terrible that there were lots of people who were emotionally invested in Poppy and Chester. We'd been telling followers how excited we were about the birth and then to have to break the devastating news to them was really hard."

It was a heartbreaking time for the family and for a long time they shied away from trying to get another of their Shire horses in foal. They just couldn't face the potential that things could go wrong. But three years later, through careful planning and making sure that absolutely everything in the labour was handled perfectly, in March 2020 Orchid gave birth to a beautiful foal. "A foal is born without any protection against infection," says David. "So it's crucially important that it gets colostrum from its mum. It's also vital that the umbilical cord between them isn't cut too quickly as they need time for the remaining blood in the mum's placenta to be transferred to the foal."

The gestation period for a Shire horse is between 10 and 12 months, so, unlike with humans, it's very difficult to accurately pinpoint when a mare will go into labour. Once Orchid's due date had come, the farmers monitored her closely during the day, and at night, Robert and David set their alarms for every two hours (Robert on the even numbers, David on the odds) so they could check in on her via CCTV. They had always thought that Orchid would give birth at night as that's when the farm is at its quietest. Mares instinctively do this as birthing at night prevents predators attacking their newborns.

On the night itself, Robert and David kept out of sight around the corner from her barn, so that Orchid could give birth naturally. They stayed very close and watched the CCTV from their phones in case of any emergencies. After a while, the foal's legs appeared, so all was going well, but as the foal was still in the amniotic sac, David nipped in like a ninja to break open the membrane, then got out of sight again. Had Orchid seen him, she may have been spooked and stopped pushing. Robert and David were determined to do absolutely everything by the book. "Our equine expert had briefed us exactly what we needed to do to make the birth go as smoothly as possible," says David, "so we followed her rules to the letter."

Taking it slowly, Orchid gave birth unaided and, although Robert and David couldn't wait to meet the new arrival, they left mother and baby in peace for the vital transfusion of blood to take place. After a few more nail-biting minutes, everyone breathed a huge sigh of relief as both Orchid and her beautiful newborn foal got to their feet, looking the picture of health. The foal was a little knock-kneed after being cooped up in the womb for so long and Robert joked that it had legs like an antique table. But the important thing was they were both fine.

Orchid had previously earned the nickname Awkward Orchid because she would kick up a fuss – and often try and kick the farrier – when her feet were touched, but she was a great mum to the little foal. Her firstborn had been named Blossom, after the Shire on Roger's childhood farm. The new foal was named Will, after Roger's lifelong friend Will Rowe who'd been such a help when Roger moved to Cannon Hall Farm.

The birth of a beautiful Shire foal was just the kind of good news that everyone needed at a time when Covid-19 was making everyone's life a misery. "Of all the projects we have ever taken on, this was the one we most wanted to work out," says Robert.

The equine expert who was instrumental in the success of Will's birth is farmer Ruth Burgess, who is often seen on the programmes made at Cannon Hall Farm. "Will's birth is the highlight of my time here. I've worked very closely with Orchid and we had carefully planned out every eventuality of her labour, so I was confident, but also very nervous on the night he was born. It was a huge sigh of relief that he was OK."

Ruth has worked for the Nicholsons since 2017 on a variety of projects, but is always happiest when she can focus her work on the Shires. "When I first arrived, I did a training plan with one of the original Shires called Lottie as she wasn't behaving. When you are working with animals that are so big, they can cause a lot of damage, and when things go wrong, they tend to go badly wrong. You have to have mutual respect; if you respect them, they'll respect you."

As the nation was in lockdown when Will was born, Robert and David had been filming reports for *Springtime on the Farm* on their phones. And, although the results were a bit rough and ready, television viewers got to see the action as it happened as Orchid gave birth. "The fact that the footage was raw and unpolished made it more lifelike," says Robert. "This is farming, after all. There's nothing staged; we just pick up the phone and film. Luckily, Daisybeck has an excellent television production team who picked out the bits they liked and made it look professional."

As episodes in the life of Cannon Hall Farm go, this was very special – a proper lift to the spirits for the Nicholson family, the farm staff and followers on social media. And, as little Will got used to being led by a harness and trained not to bite the hands that fed him, he proved he was something very special. At first, it took a while for him to pluck up the nerve to step out of the safety of his barn, but once he'd followed Orchid out on to the paddock, he seemed happier than ever.

Every day, Robert got used to visiting the young foal and giving him a special back scratch, so it was a poignant moment in spring 2020 when it was time for him to say goodbye to beautiful Will. "There were two reasons why we couldn't keep him at the farm," says Robert. "First of all, he's going to be a stallion one day and he'd be interested in all the mares at the farm so he'd be difficult for us to contain. Secondly, we decided that with a pedigree like his, there's every chance that he will make an excellent breeding stallion and pass his genes on to generations to come. So he now lives at a stud farm outside York – it's just down the motorway, so we can visit him whenever we like."

A short while later, Robert and David went to see him. "To qualify as a breeding stallion, Will needs to be at least eighteen hands and two inches tall, and well marked, with four white socks and a nice broad blaze," Robert explains. "He also has to fulfil all the breed characteristics of a Shire horse because you only breed from the very best. At the very least, I think Will should make an excellent riding horse, which I think is the future for Shire horses. Now that they are no longer used as agricultural working animals, I think we'll see them ridden more and more. As they are such calm and patient animals – not to mention sturdy and strong – they will be excellent for trail riding."

—

In March 2021, a new arrival in the cattle barn added to the continuing success story of the Anniversary Herd. A beautiful little female calf was born in the early hours one morning and David went to check in on the new arrival, filming the news for the farm's website. Armed with a bottle of iodine spray in one hand and a bowl of food in another to distract the calf's mum, he carefully approached the beautiful roan-coloured Shorthorn. "New mums are very protective and potentially quite aggressive," said David. "But it's vital that we get to newborns to spray their navels to disinfect them. The calf has obviously been pacing around, trying to feed off her mum, and their bedding has got really mucky, so I'm going to refresh it with lots of new straw so it's nice and clean and comfortable for them both, and that way the calf won't get any infection."

With lambing taking over the months of February and April at Cannon Hall Farm, it would be ideal if any new members of the Anniversary Herd could time their arrival for late spring and early summer. That way, when they are turned out to the fields to graze, conditions will be perfect for them to enjoy spring and summer on lots of lush thick green grass. On farms that primarily breed cattle, March tends to be the busiest month for calving, but it's a moo-veable feast at the Nicholsons' farm and there isn't a definitive date when Jeremy the bull will impregnate the cows.

Nevertheless, females are routinely scanned twice a year. "That way we can't miss a pregnancy and we can always catch ones that are getting close," says Robert. "We also tend to pull away and section off any pregnant cows for the last month before they are due to give birth and feed them a little less. That way the calf won't grow exponentially and be difficult to birth. The scanning also means that in the unlikely event of a cow having twins, we can mark it up accordingly and crack on with feeding the mum a bit extra."

Cows deliver their calves, provide milk for them for around three months and then come back into season. And farmers are left in no doubt when a cow is ready to mate again as she'll start to ride the other cows in the barn. It's then up to Jeremy to get to work. But it's not a foregone conclusion that every mating will result in a pregnancy, so the regular scanning is a catch-all in order for the farmers to be prepared when a cow is likely to calve.

As for the family of pigs at Cannon Hall Farm, the 45 sows are all Large White, a breed developed in Yorkshire in the 19th century, crossed with a Dutch breed called Landrace. They're bred with Hampshire boars, as the Hampshire is believed to produce the tastiest pork and bacon. The two current handsome boars are JR – named after the famous *Dallas* character – and JB – after English cricketer Jonny Bairstow. "I don't know if the original Jonny Bairstow would be very pleased if he knew there was a pig named after him," Roger says nervously . . .

Sows can grow to be enormous and produce huge litters, but farmers don't like them to get too big as they can get clumsy and there's a danger that they could roll on their newborn piglets and squash them. This happened in 2021 when the Nicholsons' biggest sow gave birth to 30 piglets, but unfortunately, only four survived. "She'd got very big and was overdue," says Roger. "I donned my long glove and when I put my hand in, I could feel that there were two piglets trying to get down the same time. Unfortunately, they were both dead and so were several more. She eventually had seven live ones, but then she rolled over and squashed another three, but I managed to nip in and lift out the remaining four and adopted them on to another sow."

This was the biggest litter size on record at Cannon Hall Farm and it was completely out of the ordinary as the average litter size for a Large White cross Landrace sow is 12 surviving piglets. Anything up to 14 piglets is comfortable for a sow to have as that's the number of teats she'll have. Sows give birth all year around with an average of 2.1 litters each year. It's a lot of numbers, but with a quick bit of maths, with or without a calculator, that means that there are around 1,100 piglets arriving every year at the farm.

Understandably, there's no such thing as a quiet day where the pig posse is concerned. At any one time, three or four sows will be pigging at more or less the same time, so the farrowing barns are always very busy – not just in spring. "It's not exactly a science – in fact, it's quite a simple thing – but we schedule everything in a notebook," says Roger. "We record the date when a sow will go to a boar, then work out three months, three weeks and three days ahead from that date. And, if they come back into season, we know that they are not pregnant, so will cross that line out and start again." Pigs don't need to be scanned as it's such a short pregnancy. "And besides," says Roger, "we know if they go back into season again because they tell us! They are very active, make a lot of noise and start trying to knock the door down."

Like any of the animals that live at Cannon Hall Farm, the farmers know what to look for when pigs are showing signs of distress. "My pig man at college used to say, if a sow looks like it needs a bit more food, give it," says Robert. "If it's losing condition, it's for a reason and it could have a belly full of piglets, and if it starts picking up the straw and throwing it around, it's a sign that she's nesting, prior to going into labour."

As the most experienced pig man of the family, Roger says that sometimes a pig giving birth is like shelling peas, or, as David describes it, "One grunt and a piglet pops out."

"Piglets are so small and sows are so big that farrowing is generally not a huge effort," says Roger. "Certainly not when you compare it to the birth of a calf or a foal. Saying that, giving birth is stressful for every animal, but it's the foundation of farming and we always try to do our very best for every mum and every newborn."

According to vet Shona Searson, it's extremely rare that vets will ever be called upon to assist a delivery of piglets, as only around two to three per cent of pigs have any difficulty pigging by themselves. Nevertheless, in spring 2021, she was called upon to help one of the Nicholsons' pregnant sows who had a prolapsed womb the size of a watermelon. As it was such an unusual story, it was featured on *Springtime on the Farm*. Having given the sow an internal examination to check she wasn't pigging imminently, Shona then administered a purse string suture – a few stitches to keep the sow's womb in place until the time when she would deliver her piglets. Luckily, she had a steady hand. "There's nothing like doing a fiddly job at the back of a three-hundred-and-fifty-kilo animal!" She laughed nervously.

Unfortunately, it wasn't all plain sailing. While Shona was hard at work at one end of the sow, Robert and David delivered a distracting meal to the pregnant porker, but, as Shona was about to finish off and just needed to inject some antibiotics into her, the sow had other ideas. Another sow snuffled in to see what was going on and all hell broke loose as the two of them had a very vocal confrontation. Luckily, Robert and David were there to restrain the patient and help finish the job in hand. "Pigs make a lot of noise, they're very strong and you have to be very careful," says Robert. "But this job needs doing, otherwise the piglets will die and we will probably lose the sow too."

With antibiotics administered, the pregnant sow was soon calm again – calm enough in fact to go on to deliver three piglets a few days later, even though her womb opening was still stitched up. Shona came back to the farm to assist with the new arrivals and, although five of them didn't make it, seven other piglets were born fit and well.

Unlike the larger farm animals, when a piglet is born, its navel isn't sprayed with iodine, as the farmers prefer to leave the sow and her young newborns to quietly bond. In any case, getting between a sow and her piglets would be a dangerous move as a sow is ultra-protective after giving birth. She might look tuckered out after birthing over a dozen children, but she'll see off anything she thinks is a predator. Another reason for not spraying the piglets is the farrowing barns at Cannon Hall Farm are regularly disinfected and thoroughly cleaned. The farrowing barns also have underfloor heating and farmers will always ensure there's lots and lots of fresh straw for the newborn piglets. So even though a single little piggy may seem inconsequential when there are so many born at our favourite farm every year, even the runt of the litter will get to live its best life there.

12

In years gone by, March would traditionally be the time when the Nicholsons would turn their attention to planting potatoes. King Edwards – or Red Kings – to be precise. Depending on the weather and ground conditions, they'd be started in March so that they'd be ready for Potato Picking Week in October. Potato Picking Week used to be a well-known feature in the farming calendar for people to earn a nice little bonus picking spuds. "When we used to plant potatoes back on the farm in Worsbrough Dale, the ladies from the local village would come and pick them," Roger explains. "There'd be a young sixteen-year-old lad in charge and this gang of ladies who'd been potato picking for years. They'd swear like pirates, so it wasn't a great introduction to the fairer sex for me, but we got through it!"

When the Nicholson family moved to their new home at Cannon Hall Farm, they went to recruit potential potato pickers, and once again Roger had a memorable encounter with a potato lady. "I was given a glass of Advocaat – you know, that horrible yellow stuff that you get at Christmas. It was the most revolting drink I'd ever tasted, but I gulped it down because I didn't want to be impolite and not finish it. But then she said, 'Ee, lad, you finished that quick, have another,' and topped me up. She was quite a character – such a character, in fact, that when she did the potato picking she wouldn't wear any pants, so it was easier when she was out in the fields and needed to have a wee."

As for planting the spuds, months before the potato 'ladies' would get to pick them, an ingenious bit of kit would be used. It was attached to the farmer's tractor and would have two bench seats on the back on either side of the potato seed container. As the tractor ploughed a furrow which the potatoes would be grown in, a bell would ring to tell each of the farm hands to alternately drop a potato down a slot. "The bell would ring every foot," says Roger. "But when it got into its stride, it didn't take long to move a foot and it went 'Ding! Ding! Ding! Ding!' You were glad when you got to the end of the row and could have a few seconds' break before it set off again."

It would take the potato-planting team at Worsbrough a couple of days to plant a 10-acre field. Bearing in mind it was a job that was generally done in March and the potato-picking machine was open to the elements, it's a toss-up whether it was the hand-numbing cold, or the incessant dinging which made the task even more onerous. But it was a job that had to be done, and it was certainly easier than brandishing a tape measure to ensure the spuds were correctly spaced out.

Spacing out the crops is basic common sense to allow roots to grow and tubers to swell underground. And when there are vast areas of land to work with, and huge quantities of crops to plant, it is even more important to space each variety correctly. At the farm where Roger grew up, turnip 'singling' was considered quite a skill, with each and every turnip spaced precisely a hoe's width apart.

"The drill planted a great long row of turnips," says Roger, "so every hoe's width, you took out the other plants in-between and left the strongest-looking plants behind. It was the skill of the farmer to know which plants to knock out of the row, so you'd be sacrificing ten plants for one. You went along with your hoe and if there were any really strong-looking plants you'd save those, but maintain the hoe's width apart as you went along each line. It was another laborious task, but it was important to get it right. Now drills are much more refined and, when we grow fodder beet, the drill will plant one seed at a time. Fodder beet seed is bigger than turnip seed, which is what makes it possible to plant it individually."

Now the potatoes and other kitchen veggies that are grown at Cannon Hall Farm are just for the Nicholsons to enjoy. It's still a bit of a sore point, but Cynthia lost her garden – her pride and joy – when the land had to be turned into a car park for visitors at the open farm. Cynthia and Roger now have a very smart area of raised beds outside the farmhouse where they grow lots of delicious veggies every year. They're rightly proud of their vegetable garden, which they tend all year around. In the spring, young shoots are nurtured in the greenhouse, and, between them, they'll decide which new crop they can try next.

And as if they don't have enough work already on their hands, the outhouses around Roger and Cynthia's garden will often be filled with orphaned or injured wildlife that Roger has happened upon along the way. He can't help himself but to give them the helping hand they need. It's something the family has always done, as Robert recalls.

"I remember one spring when we were little, our cousin Richard had been cutting crops out in the fields and he'd disturbed a hare's nest and brought us a tiny leveret that he'd found. We were all so excited because, as kids, we'd never seen a hare so close-up before and Mum and Dad said we could keep him. We hand-fed the tiny creature with a little pipette and kept it warm, but sadly it didn't make it, which was a tough lesson at the time. We'd thought that if you put enough care into looking after any animal, it would survive. I think it was our first experience of death and it was quite a shock."

Whenever they could, the brothers would rehabilitate injured and lost animals and release them back into the wild. The first lamb they helped rear was called Tiny Tim – even though it turned out to be a girl – which was kept as a breeding ewe and later gave birth to Tiny Tina. They weren't the most original names, but back then naming the Cannon Hall Farm animals wasn't quite the event it is now. There'd be racing pigeons that had gone astray, a kestrel with a broken wing, a clutch of abandoned duck eggs, the odd stray cat. Mostly, though, it was birds – and nearly always in spring.

In the early 1990s, curlews started to appear in the fields at Cannon Hall, sparking a new fascination for Robert. "There was a nest in the park and four chicks hatched and grew to be about eighteen inches tall with extraordinary long curved bills," he says. "They were incredible – like nothing I'd ever seen before. I remember thinking how lucky we were to even see them as they are flightless when they're young, so they'd be no trouble for a fox to find and kill, but amazingly all four survived. Once we established where their nests were, we put up a fence around them and hoped they'd return. And to our delight they did."

The curlew is the largest European wading bird and announces its presence every spring around the end of March. And every year Robert is delighted to see their return. "They'll disappear in winter when their young have fledged and spend the colder months in coastal areas, then they're back in spring to spend their summers nesting in the tall grass."

In the quiet countryside, away from the traffic noise and the hubbub of the public areas of Cannon Hall Farm, the song of the curlew cuts through the spring sunshine. Curlews certainly don't have the flashiest plumage – they're mottled brown so they can camouflage nicely in their nests, but with their distinctive two-note whistle, they're the town criers of spring. "To me, hearing the call of the curlew is like confirming that spring is here," says Robert. "It's like saying, OK, we're done with winter now, let's bring on the summer."

Meanwhile, in the wooded area and stream next to where Richard lives in the Gunthwaite Valley, the spring visitors include kingfishers, yellow hammers and yellow wagtails. "Mill Farm hasn't had much interference over the years in that the land hasn't been intensively farmed, so many species have been allowed to thrive. We've had some huge bats here too as they're attracted to the flying insects over the water. The most common bird we see is the jackdaw. It's smaller than a crow, but from the same family and has a silvery head. We have loads of them and they can be a pain – especially when they decide to dive down your chimney. They like to nest in crevices in buildings, holes in trees and chimneys, but luckily we have a very big fishing net!"

All three brothers share an interest in wildlife and there's always been competition amongst the three of them as to who could draw the best bird of prey. Richard generally won, as he's always been the artist of the family – so much so that if, when he was little and had invited his friends round, he'd quickly bore of them and they'd end up playing football with Robert and David. Meanwhile, he went back in the house to draw.

At their local primary school, the brothers' love of animals was further encouraged by visits from the nature bus – a minibus full of stuffed birds and other exhibits that would travel round local villages like a portable natural history museum. The boys had a teacher called Mrs Haymer who liked to teach using the "show-and-tell" approach. "She was a fabulous teacher and very matter-of-fact about the subject," says Robert. "I remember one day she said, 'Mr Armitage's hen was eating all the other hens' eggs, so he had to kill it. And when he pulled the bird's entrails out, he thought it would be interesting for us to look at them.' It sounds horrible but it was absolutely fascinating for us school kids, as you could see each stage of an egg developing – a fully formed egg ready to be laid, then a softer one, followed by a much smaller one. We were all quite matter-of-fact about it too. I don't suppose children would be allowed to see things like that now – they might need therapy!"

As well as being the family artist, Richard loved to fish. "I remember having a little fishing net and going off to find sticklebacks with Dad when I was about seven. Getting a jar of sticklebacks was considered a personal triumph back then and I remember one day I swept my net under a rock to get some sticklebacks but emerged with an unfortunate gudgeon thrashing about instead. It was an absolute monster. But, far from it putting me off for life, it made me wonder what else was lurking in the depths of the park lake, and for ages I pestered my dad to buy us a fishing rod so we could find out. Mum and Dad were on a budget, of course, and the first rod they got us was bright yellow and about six feet long. Dad and his mate Nigel took us out the first time, but none of us really knew what we were doing. We caught nothing and we got ourselves in a terrible tangle. To this day if I suggest a fishing trip, Dad will roll his eyes and say, 'No, I'll not bother. It's all just far too fiddly for me.'

"Once my love of fishing was established, Mum would add my special request to her shopping list, a pint of bronze maggots that she and Rosemary would buy on their weekly shopping trips from Clegg Brothers tackle shop on Doncaster Road. They'd brace themselves and enter with my rather grubby bait box in hand. It was another example of Mum (and Rosemary) going above and beyond for us kids, which they seemed to do all the time."

And all these years later, Richard is still in love with fishing, and has somewhat improved his technique through lots and lots of practise. "Spring is definitely the best time to go fishing, and as a lad, I spent hours by the park lake and knew it like the back of my hand. I'd set off as the sun was rising, and sometimes not return until it was setting, making sure to dodge the feeding hedgehogs on the way back across the park. One morning, I was sitting by the side of the lake when I heard a tearing sound in the turf beneath my feet. A second later, a rather confused mole poked his head up through the soil, sniffed the air carefully, and made a swift retreat. Another time I was fishing on the middle lake as a stoat danced on the opposite bank. A crowd of birds and a rabbit or two gathered to watch. Stoats use this tactic to lure in a hapless victim. Luckily for the animals watching on this occasion, the dance didn't lead to a kill. I felt privileged to be able to see it and fishing has led to countless magical experiences like this."

13

In spring, farming is very much focused on the hundreds of ewes giving birth at Cannon Hall Farm, but in March 2020 one particularly handsome ram got his 15 minutes of fame on *Springtime on the Farm*. Grizzly Bear, the beautiful Swiss Valais Blacknose ram was bought from a breeder in the Scottish Highlands for the princely sum of £2,000. He was the perfect example of his breed, with a lovely thick coat of wool, four black booties, black knees and strong curled horns.

He certainly looked a winning specimen and could have been a contender at the Great Yorkshire Show if he'd ever had the chance, but his hopes of winning prizes were nipped in the bud when he was less than two years old. His magnificent curly horns began painfully pressing into the side of his face and the raw skin there became infected. "I spotted the problem as I was shearing him," says David. "It's a smell that you don't forget and I could see that it could turn into a big problem if we didn't act quickly. The last thing you want to happen is for the smell to attract flies that lay their eggs there for maggots to develop."

There was nothing for it but to call in the vet. In a spring episode of *This Week on the Farm*, viewers watched Matt the vet cut through Bear's mighty horns using just a cheese wire and a whole lot of Yorkshire muscle. Although it certainly looked dramatic, it was a painless for Bear, although he did look a bit confused – some may say sheepish – by the end of it.

"When we've had to remove a ram's horns, it's a welfare issue and done for the good of the animal," says David. "It means that we'll never be able to exhibit him as an example of his breed, but his strong pedigree is in his genes, so we're hoping he will father many more Swiss Valais lambs in the future. He's fathered loads of great lambs already, including Brussel and Sprout, and the quality of the markings on our Swiss Valais lambs has improved no end. We love the breed so much that we're looking to go back and buy another ram from the breeder in Scotland so that we have two bloodlines on the farm. That way we can sell unrelated pairs of males and females on to other people. Swiss Valais sheep are always bred as pedigrees and are far too valuable to be in the food chain, so they will always be sold on as pets, or to other breeders."

Talking of future generations, every breeding male on the farm needs to have his faculties checked before he is put to work with the females. In spring 2021, it was time to see how Ted, the young Highland bull, was likely to perform when breeding time came around. After guiding the 20-month-old into the cattle crush, the width of his testicles was measured and his sperm was quality tested to ensure he could deliver the goods. It was good news as Shona the vet confirmed that Ted should be able to father top-quality Highland calves in the future. And it wouldn't be too long until the farmers found out for sure that he could.

A bull with urges can be a bit of a handful. "Once we breed a male calf, his calm nature changes," says David. "The risk is that he'll want to have his way with every cow that he sees and that means he's much harder to handle."

It sounds a bit too basic to be the reliable truth, but the size of an animal's testicles really does indicate whether or not he's likely to be able to produce enough sperm to father offspring. It might not be a scientific test but it's a useful gauge when farmers are investing in a new breeding male. In March 2021, Robert and David employed the same method in their search for a new Kerry Hill ram, a distinctive panda-eyed black sheep with black ears, mouth and nose. Up until early spring 2021, their handsome ram Spartacus had been the head of the Kerry Hill flock at Cannon Hall Farm. When he died, Robert and David took a road trip to a farm in Whitby in North Yorkshire to find his replacement. They picked out a prime specimen and gave him the vital testicle once-over. No tape measure this time, just the advice that Robert had been given. "I've been told they need to feel like strong biceps," said Robert, so, with that benchmark, he gave the ram's tackle a gentle squeeze. And finding they were indeed like biceps, it was job done. The financials were sorted and it was back to Cannon Hall Farm for Romeo the ram.

By the following spring, Robert and David would know if he'd lived up to his name . . .

—

As every celebrity fan will know, March is the month of Hollywood's annual Academy Awards, and, while Robert and David might not be ready to collect any Oscars for acting, they're no strangers to picking up prizes. In March 2021, Daisybeck won the award for Best Lockdown Programme by *Broadcast* magazine for *Springtime on the Farm*. And there were more nominations for awards to follow from other magazines and newspapers, the farming industry, the tourism sector and official retail bodies. Due to the pandemic, many award shows were held virtually, and several times members of the Nicholson family found themselves in their best bib and tucker, sitting in front of their computers, and joining other nominees in the same situation as the host presented the awards. It's not quite the same as rubbing shoulders with the stars at a swanky hotel. But then again, even when you attend the fanciest award shows in the flesh, things don't always go to plan, as David recalls . . .

"One spring, we were all going off to an awards night in Birmingham," he says. "Robert and I were sharing a hotel room close to where it was taking place, so we checked in to the hotel and hung up our suits. Unfortunately, Robert had assumed it would be the usual posh dress code, so he had mistakenly packed his tuxedo instead of just a lounge suit (which was stated on the invitation – he just hadn't bothered to read it!). Well, his teddy got well and truly thrown out of the pram. He stuck his bottom lip out, got a right monk on and said, 'I'm not going then. I'm not going to wear a bloody penguin suit; I'll look a right chump!' And no matter how much I tried to convince him that he'd be OK to go in his formal black tie, he wouldn't have it. I wouldn't laugh at him, I promised (well, not too much). Finally I said, 'Look, we've got an hour before we need to be there; let's go out and buy you a suit. We were in Birmingham after all, so there was bound to be plenty of suit shops in the vicinity.' We ordered a taxi to the nearest shopping centre and off we went. But, of course, sod's law, we got caught in the most horrendous traffic jam. We just weren't budging and all the while the clock was ticking. Meanwhile, Robert was getting more and more sulky, but I was determined we'd make it. All of a sudden, we saw this factory outlet shop. We jumped out of the car, rushed in, Robert pointed at the suit in the window, said what size he was after and off he went to try it on. What can I say? He went from being the grumpiest I've ever seen him to looking like the cat that got the cream. He was all smiles, showing off his little waistcoat and his new grey checked suit. It had a matching pale mauve shirt and purple tie, so he bought the lot and off we went to the awards do, with Robert's work clothes in a carrier bag.

"We managed to get to the venue really quickly and in the end we were only ten minutes late. He was still peeling off the price tags as we got to our table. We didn't win anything, but we all had a great time and Robert was cock-a-hoop in his new whistle and flute. So it was certainly a night to remember."

Awards nights are a great way for many of the team members to come together and celebrate various achievements on Cannon Hall Farm. Whether it's for excellence in farming itself, tourism or farm shop produce, everyone likes to have a pat on the back and it's the time to hang up the overalls for a night. "It's a great boost for our teams to come together for awards nights," says David. "Especially during spring when the weather might be lousy and everyone is feeling a bit flat. Some of our staff members have worked with us for so long that they are like family now and it's a great excuse to let their hair down. We've had some mad nights . . ."

—

In late March and early April, it's spring cultivation time – often just as the farm is also gearing up for Easter and the busiest period of lambing. "Sowing certain crops, along with lambing, is one of those non-negotiable spring jobs," says Robert, "and you have to work with nature, so it has to be done at a certain time. You have that one window of weather, where suddenly you wish you could clone yourself to get everything done."

"The first chance that you get to jump on your spring cultivation, you do it," he says. "If you can get it in early, it makes such a difference to the crop. It would be nice if the farming calendar worked in a way that you could calmly go from one completed job to another, but spring cultivation and lambing time have always clashed. However, we're lucky in a way because we no longer plant as many crops as we used to. Our fields are pretty much all grass now and we've adapted the system and plant our stubble turnips and fodder beet in June. The sheep and cattle then have something to graze on through autumn and winter."

Even though there's always more than enough work to do on the farm, Robert misses working out in the fields. "It was always an enjoyable job to do. You'd be off on your tractor pulling a harrow and that would work up your seed bed and plant your crop. It's quite therapeutic really, getting your lines nice and straight and seeing your progress as you work across the field. For someone with an obsessive-compulsive disorder like me, I see it as a strength. I just like things to be right."

To get the maximum results from a field, whether it's for grazing or for planting crops, a good dose of fertiliser is also in order. "If you fertilise a field, you'll get just that extra bit of growth on it," says Robert. "In the old days, we would all have had a go with the tractor and the spreader. You know when you've got it wrong because all your grass is stripy – and, if it's near a main road, it's really embarrassing. Now we get a contractor to do our fertilising for us. We leave it to the professionals."

In the days when Robert used to take himself off to do the cultivating, he'd enjoy a bit of wildlife spotting along the way. "Even though you are sitting in the tractor cab, you get to see the world around you and take everything in at this lovely slow pace. I'd have the wireless on and listen to the cricket, taking in the curlews and the lapwings flying around. As they may have a nesting site in the field you'd always have to be mindful not to destroy them. I'd usually have a good walk around the field first before I started working on it, but many a time I've had to roll around a lapwing nest so as not to disturb it. But if it's early in the year, they won't have had time to lay their eggs, so I can keep my lines lovely and straight, and it's ready for the lapwings when they want to visit."

14

The end of March sees several significant anniversaries in the history of Cannon Hall Farm. First up, on 18 March 1989, young sweethearts Robert Nicholson and Julie King married, a year and five months after meeting. "It was a radical decision as we were both so young," says Robert. "I was just twenty, so my going-out days hadn't been over long, but it just seemed the right thing to do. We got married on the eighteenth, had a couple of nights' honeymoon in Newcastle, then came back to Barnsley as the farm was opening to the public the following week. We were certain that we wanted to be married, but it was a time of real uncertainty for the family because at that stage we didn't know whether we'd be able to stay at Cannon Hall Farm. The family was converting farm buildings so that we could open to the public, but everything was being done on a shoe-string budget because we just didn't know whether the open farm idea would work out or not." So, as soon as the newly-weds returned from their mini-moon, Julie was straight back to work at the tea room, and Robert helped Roger, Richard and David with the final preparations for the grand opening.

And from openings to closings. In 2020, also on 18 March, the family decided to shut the farm to the public due to

Covid-19. It was certainly momentous as it was the same day that Robert and Julie's granddaughter Nelly was born, so all on the same date, they had a wedding anniversary, a birth and the closure of the family business. "It was such a worrying stressful time," says Robert. "Katie was in hospital as she had terrible complications throughout her pregnancy and she was very sick. Suffice to say it was a day of very mixed emotions – but overriding it were happy emotions. We were able to visit Katie and have a cuddle with the new baby, which, in hindsight, was very lucky, as soon tougher restrictions were brought in. After that, we could only wave at her through the window. It was really tough for all of us – but especially for Katie and Julie."

Now, when 18 March comes along, there will be happy memories for the Nicholsons, with little Nelly being the apple of her parents', grandparents' and great-grandparents' eyes. However, 24 March . . . that's the most important date in Cannon Hall Farm history as it's the day the Nicholsons' farm was opened to the public for the very first time.

"I remember that day vividly," says David, "as it was a mad dash to get everything ready. We've always prided ourselves on the fact that the farm is spotlessly clean, so, as usual, all the animals were mucked out and the farmyard was swept nice and early. But we were still putting up signs right up to opening time and scratching our heads whether anyone would actually turn up to visit. Then, at half past ten, we officially opened. We'd run an advert in the *Barnsley Chronicle* newspaper about opening to the public, but we hadn't put the admission charge on it, so I could see some people turning up and saying, 'Why would I want to pay a pound to see some bloody sheep? I can see them out in the fields for free.' The trouble was, next door, the gardens at Cannon Hall Park were absolutely magnificent with all of their spring flowers out in bloom, and it was free to visit there, so you can understand why they were reluctant to shell out to visit us. You could have bought a choc ice for the same price!

"But, all in all, it wasn't bad for a first day, and we got a few people through the gates. Later, we celebrated with a few glasses of beer and pizza from the local Italian restaurant. I remember my dad saying, 'We're not going to have much of a business if you keep spending all the profits on pizza!' Luckily, we had more people on the next day, more again on Easter Saturday and Sunday, and, by Monday, word had really started to get round, and lots of people were turning up to visit. We even got to the stage where we were taking good sums of money on the door. I remember we used to have code words for how much money we'd earned at the entrance gate. A 'dwarf Lop' was £15, a 'Kunekune pig' was £250 and a 'llama' was £1,000. So, if someone said we'd 'made a llama', that was quite a result!"

With every new business, there are often teething troubles, and, in the first few years of Cannon Hall Farm being opened to the public, there was one particularly dramatic challenge. "One spring, a coach dropped off a school party and decided to park near the houses as we didn't have a proper coach park at that time," David explains. "There wasn't much room and, as the driver swung the coach round, it collided with an electric pylon and completely took it down. Luckily, the driver was unhurt, and he climbed out of the coach and up on to some breeze blocks to get out of the way. To steady himself, he grabbed hold of the electric cables . . . Unbeknownst to me, the power had already been cut but I remember it almost in slow motion – running towards him saying, 'Noooooooooo!' Of course he was OK, but, unluckily for everyone else, the power had gone all over the farm. It had also cut off the nearby pipe factory and the whole of Cawthorne Village. Needless to say we weren't too popular with our neighbours that spring . . ."

15

As late March gives way to early April, it's as if a page has been turned on the continuing chapter of spring. All the light that everyone craved back in January now floods every space, as if the dial has been notched up to at least seven on the brightness scale. There are still brighter days to come, but there's so much more optimism in the air than in those dank long winter days. Out in the fields, hares are busy at work trying to pair up and breed, with a female boxing off any unwelcome advances until she is happy she has found the right mate.

"We used to see lots of March hares in the woods – all dashing about in spring," says Cynthia. "We see fewer pairs now, which is a shame, but still they're always exciting to see." Unfortunately, farming methods have changed so much that some animal species have all but been wiped out.

"There was one time when we used to see grey partridges absolutely everywhere too," says Roger, "but the grass is cut so early when certain birds would still be nesting that it has all but obliterated them. They need an arable area and they haven't adapted to the change in cropping methods that farmers use to manage their land. We also used to have lots of skylarks, which look like quite insignificant small brown birds until you see

Roger and Cynthia on their wedding day, 2 October 1965

Roger on his tractor with farmhand Brian Machin

Roger and Cynthia enjoying a
night out in Huddersfield

Baby David

Roger aged 20, turning out for
Cawthorne Cricket Club

Robert and Richard in the early years in the farmhouse

Roger with David, Richard and Robert

David, Robert and Richard's first photo shoot

David, Robert and Richard
with grandma Olive Dickin

Family photograph, back
row: Auntie Flo; Roger;
Richard; Roger's mum, Rene.
Front row: David; Cynthia;
Roger's sister Olive; Robert;
Olive's husband, Vinny

Roger, David, Richard, Cynthia and Robert

As a young teenager, Robert was Yorkshire ranked at table tennis!

David and Robert planting oak trees to create a wildlife area; photograph taken by Richard

Richard, Robert and David in the farmyard at Cannon Hall Farm

Robert and Julie getting married on 18 March 1989, the week before the farm opened to the public for the first time

Roger with border collie Lassie and her pups, including Flossie, who eventually became the dog on the farm's logo

OUR FAMILY AND FARM TODAY!

Robert, David and Richard while helping out at the farm's pumpkin festival

Cynthia and Roger enjoying a holiday in the Cotswolds

Drama at the farm as vets deliver a Highland calf
by caesarean in front of a crowd of onlookers

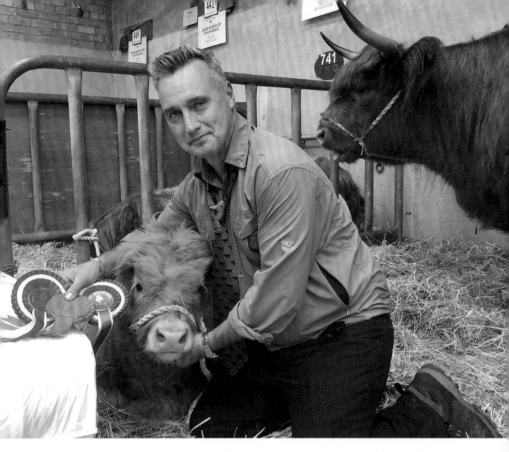

David with mum Fern and her calf Ted, winners of the Cow and Calf class and
reserve champions in the Highland class at the Great Yorkshire Show

Robert, David and Roger outside the White Bull restaurant

Roger has loved his farming life, despite all its challenges

David and Robert checking on the ryegrass

Spring 2013; heavy snowfall means the
sheep and lambs need extra food

Roger discovers a lamb buried in a snowdrift suffering from
hypothermia. Robert carries it to the Land Rover. They have
to work fast to get it back to the farm and warm and dry

Sheep and lambs running through the snow to reach
newly delivered feed supplies

This toughest of springs was hard work for everyone
on the farm and the animals too!

Triplet lambs safely delivered to this Texel ewe

Springtime is our favourite time of year at the farm, with new life at every turn

Tiny Lamb became a social media favourite as Robert broadcast from the lambing barn in 2020

Hundreds of lambs are born on the farm each springtime

David with twin pygmy goats

An alpaca with her cria

A pygmy goat kid

The piglets and sows have underfloor heating at the farm, to help keep them warm

David with a Shetland pony foal

Ozzy Horsebourne

Jon Bon Pony

Jeremy the Shorthorn bull
is a big favourite at the farm

Fern, our prize-winning
Highland cow

Mary from the Dairy, who sadly
passed away in 2021

Pongo the Dutch spotted sheep
was another lockdown favourite

them spectacularly raising up from the ground as they sing. They also nest in long vegetation, so there just isn't the habitat to tempt them here now.

"It saddens me that aggressive modern farming has put so many species at risk. As a family, we're committed to doing our bit to help protect wildlife, and accordingly we have adjusted certain ways that we farm. I believe that we should concentrate on improving the lot of our present native wildlife rather than trying to turn back the clock and reintroduce certain animals. Britain lacks areas of genuine wilderness with not a great deal of distance between our towns and cities, and therefore it's unsuitable for releasing large animals and birds higher up the animal food chain that will cause a sudden imbalance. It would seem common sense to improve what we have got and not to interfere with nature – like our forebears did with the introduction of grey squirrels and mink, which proved such a problem to our native species.

"Also, flowers do seem to come up much earlier than they used to," Roger continues, "and that's all down to global warming, I'm afraid. It's lovely to see more colour everywhere, but climate change upsets the balance for wildlife. When the seasons change unpredictably, it throws things out of kilter. Plants may grow more quickly, but where are the pollinators that feed from them? And, if they're not around, there won't be food for the birds to survive."

It's hoped that the creation of new wildlife corridors in Britain will encourage natural habitats that are rich in wildlife to link together, ensuring more native breeds will be able to continue and flourish.

Aiming to encourage more native insects and butterflies back to Cannon Hall Farm, in spring 2021, Roger cornered off a plot of one of his fields to create a wildflower meadow. It's next to Tower Cottage, where Roger and Cynthia first lived when they were married, and is now home to their lifelong friends, Rose-mary and Nigel Brain. "I persuaded Roger to give me a little corner," says Rosemary with a wink. "Obviously it's all valuable grazing land for his sheep, but I knew it would be spectacular when it was all in bloom, and wonderful for the bees and but-terflies. It's a mixture of poppies, cowslips, lilac scabiosa and white daisies. As a result, we've had lots of butterflies – I've counted sixteen different varieties and all manner of garden birds, including siskins, lesser redpolls and linnets."

Cynthia and Roger's garden is also in full spring bloom by the end of April and the planters that line the entrance to the farmhouse are awash with colour. Every year, Cynthia adds a few more bulbs to the pots and, come rain or shine, they never disappoint. And more flowers bring more winged spring visi-tors, such as busy bees and the unmistakable curlew.

Although they don't rank as the nation's favourite bird, the arrival of the cuckoo is another avian sign that spring has sprung. "They can be quite aggressive things," says Cynthia. "They'll chase off other birds – even buzzards – and lay their eggs in other birds' nests. There's an old saying that, if you heard the call of the cuckoo, you should turn your money over because the cuckoo call would increase it." Another custom suggests that when hearing the first cuckoo call in spring, one should run three times in a circle to ensure good luck for the rest of the year.

It's good news all round when the Cannon Hall Farm bees start visiting from spring onwards – not least for local honey producer Phil Askham, who collects the bees' spoils and sells his honey in the farm shop and other select stores around York-shire. It comes from various hives in the area including the dozen hives in Deffer Wood that overlook the stunning South Yorkshire countryside. Lilacs bloom there in spring and the forest floor becomes a carpet of bluebells. "On a lovely sunny day, the smell of the bluebells and the sound of the bees is one of life's greatest pleasures," says Roger.

Retired beekeeper Phil Askham has sold his own brand of Yorkshire Honey at Cannon Hall Farm since the early 1990s and the company is now run by his business partner, Pat Foster. In the seventies, Phil bought his first bee colony after seeing an advertisement in the paper and, like the colony itself, his business grew. "I've always grown my own food," Phil says. "I have a large plot of land and I grew fruit and vegetables and kept chickens and pigs, so it was like *The Good Life*." And once the local health food shops and farm shops got a taste of his amber nectar, soon they wanted more. "We have three varieties of honey that we produce," says Phil. "Our Spring Honey is from all of the spring flowers, such as apple and pear blossom, oilseed rape, sycamore and gooseberries. Our Summer Honey has blackberry, willow herb, clover, linden and wild flowers, with the addition of privet, horse chestnut and other summer flowers. Then, at the end of July, we move the hives up to the Moors, where the bees collect ling heather for our Heather Honey. It's what I consider to be the connoisseur's honey and it's very much sought after."

Bees tend to stick to a three-mile radius of their hives, so you can be sure that Phil's Spring and Summer Honeys will be bursting with goodness from Cannon Hall Farm. Phil has around 180 hives all around the South Yorkshire area, where there are rich pickings for honey collectors. "There are lovely mature lime trees and yew at Cannon Hall Farm, plus oaks and ash, pastureland for clover and hedgerows, which are great for brambles," Phil explains. "The neighbouring arable crops are wonderful too and the bees love oilseed rape – they're on it like the fat kid on a Rollo! The traditional oilseed rape crop used to drip with nectar. There are also lots of country estates in the area, such as Wentworth Woodhouse and Wharncliffe and disused industrial areas, muck stacks and canals that have all rewilded beautifully."

16

As much as the team try and plan for the main lambing season to coincide with Easter, nature doesn't always play ball. It's not a precise science because when the rams are put in the fields with the ewes to mate, they can sometimes seem more interested in the grass than the fertile females. Consequently, they can take a few days to tackle the matter in hand. Even the most virile ram likes to take things in his stride sometimes. We may think that the sight of a flock of fabulous ewes would be the animal equivalent of the Playboy Mansion to a randy ram, but you can lead a horse to water . . .

"We'd love it for lambing to start just as the visitors are coming through the gates on the first day of the Easter holidays," says Robert. "For each breed to happily lamb with the least intervention on our part. Then it would be nice if the ewes stopped delivering at the end of the afternoon so we could all have the night off. But, of course, it doesn't work out that way. Equally, though, we always have our fair share of early surprises."

Late in March 2021, farmer Dale was filming reports for Cannon Hall Farm's website at the time, and quickly turned the camera on the action. "The Zwartble tup must have been on it like a car bonnet," said Dale. "We've had loads of this particular breed born and it looks like it's going to be a busy day."

With farmer Charlotte at the business end of the Zwartble ewe, it was clear to see that the twins she was carrying were going to be difficult to deliver. She carefully manipulated the opening to the ewe's womb and felt around inside, trying to find out why the ewe was struggling. "It's all there, she's just taking a bit of time to get going," said Charlotte. "I don't want her to get too tired and stop pushing." After positioning the first lamb's head and little feet into the perfect delivery position, it was then a matter of waiting for the ewe's next contraction so that she could time a good firm pull to release the lamb.

As soon as the lamb took its first breath, it was obvious what the problem had been – she had a very big head and a sizeable body too. Unlike with human births, newborns aren't put on the scales and whether they are a boy or a girl doesn't matter at that stage either. It's just a case of getting the ewe to start licking the lamb to form the natural mother-baby bond. Newborns will then instinctively start feeding. On this occasion, however, there was a problem as the mum seemed so preoccupied with licking her baby clean, the thirsty lamb couldn't get underneath her and reach her teats to take any milk. This first feed is essential for any lamb's survival and there was a danger that the newborn would start wandering off in search of milk from another ewe. Not only that, its sibling was on its way.

The farmers decided to move the ewe and her lamb into a separate pen in the hope that the natural order of events would take place – she'd let her first lamb feed and then give birth to its twin, but once again Charlotte had to give her a helping hand with the delivery. Out popped number two – with an equally big bonce. Charlotte sprayed its navel, the ewe started cleaning the newborn, and both lambs finally had their first feed. As farmer Dave might say, it was happy days all round.

A day later and it was all go in the Roundhouse as farmer Ruth found herself delivering two sets of early Zwartble triplets on different ewes at exactly the same time – talk about multi-tasking! What made it even worse was the first lamb on its way was breech. If Ruth hadn't stepped in to help the delivery, the lamb would probably not have survived as its umbilical cord could break while it was still in the womb. It would therefore drown inside its mum. But with the help of Ruth's magic hands, the first lamb was born swiftly and safely and was soon enjoying a lick from its mum.

Before she'd had time to catch her breath, Ruth helped deliver the first triplet from the other Zwartble in trouble. This time the lamb was doing a one-armed dive out of the womb, with one of its front legs tucked back. This would have meant a hefty blockage for the struggling mum to contend with. "I pulled the other leg forward and just left her be," said Ruth calmly, before the next drama – the ewe was trying to give birth to its other two lambs at the same time. By carefully feeling inside the lamb to sort out the jumble, Ruth was able to deliver the first lamb, then be on hand to deliver the second as it was another breech. This was not going to be an easy day.

Knowing how fast she had to work, Ruth quickly pulled out the breech lamb and immediately gave it a quick swing back and forth to clear its lungs. "The biggest of the three lambs was coming out fast – and coming out backwards, so we could have been in big trouble and that's why I had to step in." Moments later, though, you'd never have known there'd been any dramas as all three lambs were happily bleating "hello" to their mum.

Meanwhile, back with the first patient, the second of its triplets was delivered easily but number three was reluctant to make an appearance. This time its head was over to one side and Ruth needed to manipulate and straighten it up before giving a helping hand to bring it into the world. With a little sneeze, the lamb took its first breath and Ruth placed it next to its equally lanky-legged siblings. "You can never guarantee that they are going to come out the way you want them to," says Ruth. "They can have their legs back or their head back or just be waiting in there way too long. But, on this occasion, we had six lambs born in less than twenty minutes, so it can be quite a challenge."

It doesn't matter what breed of sheep it is, any ewe can have trouble giving birth sometimes. Coincidentally the very first time that farmer Kate had to assist with a breech birth, it was also a Zwartble. "It was in my first year here," says Kate, "and normally I would have just called David to help me with a breech, but I knew I had to act fast as the lamb was already halfway out. Its back legs were hanging out so I rushed over and just pulled it quickly. It was really gasping for air and I had to do a lot of cleaning around its face to make sure its airways were clear. All the time my heart was just hammering and I was thinking, 'Please stay alive, please stay alive.' Before that day I'd always just been at the front end of the ewe, stroking its head and saying, 'Good girl, good girl.' So it felt particularly wonderful to have been able to save this one."

There were multiple births left, right and centre at Cannon Hall Farm in April 2020. David was in the right place at the right time when he was assisting a triplet Boer goat, then its sibling quad arrived moments later. What's more, the fourth kid was born in the fluid sac and would have died if David hadn't been there to break open the sac for the newborn to breathe. Having had four kids, the nanny goat was bewildered and tired and the smaller of the quads was struggling to drink, so Robert came to the rescue with a feeding tube. By inserting the tube straight into the tiny kid's stomach, he was able to give it some colostrum so that it would have the strength to start suckling on its own. The new family was then transported to the Roundhouse to get some intensive care with regular hand-feeding and comfort from a heat lamp.

—

In April 2021, the farmers decided to let all of the Herdwick ewes give birth to their lambs out in the fields as it had been such a mild spring. As traditionally they were reared in the Cumbrian hills, Herdwicks have no trouble lambing outdoors. The weather held up and the first of the new batch of lambs started to appear. Farmers Alex and Tom took to social media in late March to explain how it was all going. "This is the first time that we've lambed them outside," said Tom. "and it's good for the sheep, but the problem is it's not always easy to find the newborns. We need to move them from the Picasso Field, where they give birth, down to a quieter paddock." It wasn't the easiest job as the Picasso Field is up the top of the farm and the lambing paddock is down near the horses' sand paddocks.

Once they'd found one particular missing lamb, it was a matter of Tom scooping it up and getting its mum to follow them down through the fields. With the help of his uncannily accurate impression of a newborn lamb bleating, he was able to encourage the ewe to keep following him. Meanwhile, Alex tried to stifle his laughter as he filmed the delicate operation. Then, once they reached the paddock, Tom quickly sprayed the newborn lamb's navel with iodine to stop the umbilical cord wound getting infected, then off it went, bleating (the lamb, not Tom this time) to find its mum. Job done. Another happy ending on the farm.

PART FOUR
April

Tasks for April
Easter lambing
Fertilising, final drilling and spring cultivation
Other scheduled calving, kidding, farrowing and animal births
Hand-feeding and nurturing newborns
Tree felling and woodland planting
Fitting and checking nesting boxes for swallows

17

In Christian calendars, the first full moon of spring is called the Paschal Full Moon and Easter is always celebrated on the Sunday after this occurs. The date that the Paschal Full Moon appears is calculated by the lunar calendar, which has fewer days than a solar year – a lunar calendar being approximately 354 days. But even though there's only 11 days' difference between the lunar and the solar year, the date Easter falls on every year can vary enormously, from as early as 22 March to as late as 25 April. That poor Easter bunny doesn't know if he's coming or going.

The Easter period is the busiest time at Cannon Hall Farm in terms of the number of visitors, demands on the shop for extra-special seasonal treats, plus the amount of extra work with lambing, so the team need to be even more organised than usual. Things may have seemed busy in February when the first lot of lambing took place during the February school half-term holidays, but that's just a warm-up for the main event. Each year, approximately 800 lambs are born over Easter, plus around 100 goat kids as well as the arrivals of Shorthorn calves, multiple litters of piglets and other breeds around the farm. If you factor in all of the other small mammals on the farm and the residents of the Reptile House, it's a mind-boggling number of new lives.

In addition to all of the animal arrivals at Cannon Hall Farm, it's also the time when the farm celebrates Easter with special events for all of the family. As long as there isn't a pandemic to contend with, there'll be egg hunts, ferret and sheep racing (genuine woolly jumpers!), tractor trailer rides, animal talks and more. "Easter is our busiest time for visitors," says marketing manager Nicky Hyde, who has worked at the Nicholsons' farm since 2015. "One time we made our Easter event extra special by having one of Richard's artist friends design an 'egg machine' where children could claim a Creme Egg. For the first day it worked amazingly well; however, so many children came to use it that it couldn't keep up, so our team had to resort to the good old 'string and push tactic' to deliver each egg.

"Another time I ordered an animatronic dinosaur for an event – we waited months for it to be delivered, but when it arrived, it became very clear that a very specific body type and level of fitness were needed to be able to wear it. We got an actor to wear the costume and act out in it but he barely managed one session before saying he couldn't continue because it was so heavy. Eventually it was a choice of either not having the dinosaur or finding someone to go in it. Hey presto, farmer Dave came to the rescue! He climbed in that heavy dinosaur suit and did the full day without complaint. I think he enjoyed it, actually!

"Events at Cannon Hall Farm have gone through quite a process of evolution and there have definitely been some learning curves along the way. Luckily, with the farm being so family-oriented, and many staff having children of an age to be interested in the farm, we are never short of ideas. However, things have been very make do and mend at times. During the five years I've worked at Cannon Hall Farm, it's certainly been action-packed: from outdoor cinemas to music tribute events, from Easter to elves. The key point is that each year we try to do better, be more creative and what I love about Cannon Hall Farm is that no idea is ever too crazy. No matter how absolutely bonkers I think I sound when I come up with some of the ideas, I'm always thrilled when the family get behind it and say, 'Yes! Let's do it!' "

Before the expertise of Nicky Hyde came along, the first Easter activities that were held on Cannon Hall Farm were not quite as professionally planned or executed as the ones visitors can now enjoy. When the farm first opened up the public, it was very much a case of trying things out to see what would stick.

As David explains, "One year we got this idea of doing a big Easter egg hunt and we got some giant camouflage nets, spread them over a field and made it like an army assault course, with branches to hide the eggs under and lots of muddy puddles for kids to jump in. The idea was that if you found a rubber egg, you could swap it for a chocolate treat. Now, it all sounded all right on paper, but when the day for the egg hunt came, it was so busy that all these kids piled in to find the rubber eggs, swapped each one for some chocolate then dived back under the camouflage straight away to find another. It was just utter chaos. I ended up with a bucket of both chocolate and rubber eggs and hundreds of children chasing after me. I felt like the Pied Piper. In the end, I just chucked the whole bucket up into the air and there was this mad scramble to pick up all the chocolate. I think we all agreed that it didn't really work out that well. The children all enjoyed it, but it certainly wasn't slick. Put it this way, it wasn't exactly Disneyland."

Even before the farm was opened to the public, Easter was always a busy time for the Nicholsons – and not just because of lambing. "When we were kids, every spring Dad would rent a field out to the Caravan Club and we'd have various caravan rallies that would pitch up for a weekend or a few days," says Robert. "It was great fun. All of a sudden we'd get this new influx of kids to play football with and share our dens and tunnels. It was like a whole new set of friends just appeared every weekend. Although the facilities were very basic for the people coming to the caravan rallies – certainly compared to what's on offer at caravan sites these days, it always seemed to work really well – especially when the weather was great. We did it from when I was about eight years old right up to when we opened the farm to the public, so it was obviously a viable business for quite a few years. Dad was always thinking up new ways for how he could make a bit more money on the farm and this was definitely one of his better ideas!"

During those years when the brothers were all old enough to play out in the fields without an adult being with them, yet still too young to be helping out with the farm jobs, during lambing time they'd be off on their own mini Easter adventures. Fishing always figured highly, and there were plenty of opportunities (legal or otherwise) to try and land a catch.

When their own children were growing up, they would try and hatch new baby chicks that they could pet in the Easter holidays. They'd collect the eggs from Roger's bantams and carefully put them in the incubator, marking each one with a T for Tom, K for Katie or P for Poppy (this was before their cousin Marshall came along) and they'd have a mini competition to see whose egg hatched first. The family competitive spirit is obviously in the genes.

The Nicholson brothers always had a cheeky scheme up their sleeves to make some pocket money, their favourite money-maker being taking a baby chick to Cannon Hall Park and charging a penny to stroke it. "David would hold the chicken, 'cos he was the cutest," says Robert. Or they'd be selling boxes of random eggs (with questionable freshness) that they would find in and around the barns. "I now think there's nothing more lovely than seeing a hen coming out of a hedgerow with half a dozen chicks behind her," says Robert. "But when we were little, we'd be thinking, 'How did we miss those eggs?' We were constantly on the lookout for them. Every egg that we found had the potential to make us a bit of money."

There's a saying "where there's muck, there's brass" and, in the spring, the brothers found the ultimate way to make a bit of cash out of any old rubbish. "We had a good pal called Jim who worked over at the museum at Cannon Hall," says David. "He was an ex-professional boxer and he used to get us to collect sheep muck for his tomatoes. I don't think we ever had any gloves, mucky beggars; we'd just go and hand-pick a great big bag load for him and he'd give us something for our troubles. He was quite picky, mind – he didn't want any of the straw or the bedding in it, just pure muck. Other kids might be off somewhere nice for their school holidays, but that's how we spent ours! That's how glamorous it got in our day!"

Like many dates in the calendar, which are now a much bigger deal than they used to be, Easter used to be just another day at Cannon Hall Farm. "We'd get an Easter egg each – but nothing fancy," says Robert. "Just a Dairy Milk or Cadbury's Buttons. Our hearts would sink if we saw that it was a dark chocolate one – but we'd still eat it anyway."

Now with the demands of lambing upon them, there's little time for chocolate, or a great big family get together. "A good Easter on the farm makes a good year," says Robert. "In fact, when we were first looking at opening up to the public, we figured out that we had these twelve weeks of opportunity during the school holiday year when we could encourage as many visitors as we could. So we always try and ensure that we have as much lambing as possible during the February half-term and Easter holidays to make it a really special time for families to enjoy the spring. And nothing shouts spring more than new lambs being born."

For as long as they can all remember, as soon as they were old enough, the Nicholson brothers have always helped out at lambing time. "If my dad was about to lamb a sheep, he'd get the lamb in the right position and then get one of us to help pull it out," says Robert. "We weren't doing any of the manipulating, or any of the real work, it was just Dad's way of letting us feel like we were involved. And I never minded about the gluey mess or any of the other stuff, I just wanted to get stuck in. I was always determined to be the best at any farming skill, best lamber, best everything really!"

If you've not got the stomach for all manner of bodily fluids and challenging solids that an animal produces, then being at the business end of a ewe during lambing isn't for you. "They always come out really gunky," says farmer Kate. "But it's a really nice smell – it's fresh and sweet, and the lambs are so warm because they've been inside their mums for so long. I'll get home after work and my waterproofs will be covered in powdered milk, colostrum, afterbirth and all sorts of bodily fluids and I'll be told that I smell awful, but I won't be able to smell anything at that point!"

During lambing time, it's a case of just getting stuck in and getting on with the job. "Sometimes you can have a problem like ring womb," Kate explains, "where the ewe's cervix isn't dilated so you have to manually assist it to open up more, first with one finger, and then with two until you can get your whole hand into the womb, then you can work out what's going on with the lambs that need to be delivered. Or sometimes you have two lambs inside and you have to tell which leg belongs to which. Meanwhile, they can also twist around each other. I always feel for the nose, the forehead and the mouth, then the two feet and eventually you can tell the difference between the front legs and the back legs. You learn how it all feels just by doing it all the time. It's a bit like doing a 3D jigsaw puzzle – but doing it in the dark."

As experienced as Kate now is, she realises when she needs to let the experts take over. "One time during the filming of *Springtime on the Farm*, I found myself assisting a really tricky lambing. I didn't want to interrupt Rob and Dave when they were filming, but it got to the stage where I had to get help from Dave, who's extremely experienced at lambing, then he needed help from Julian Norton, who's an extremely experienced vet. And even then it was a difficult delivery. You can never be too proud about asking for help – especially when there's a life in the balance. I'd hate to think I didn't call someone and I lost a lamb simply because I thought I could do it myself."

The team at Cannon Hall Farm are so experienced with lambing that they can generally sort out any tricky delivery problems and also identify diseases and know how to treat them without having to call out a vet. For example, if a ewe is suffering from Twin Lamb disease when her calcium levels suddenly drop, or there are signs that a sheep may have a disease like coccidiosis, orf or mastitis.

"In the last couple of years, we've been able to administer anti-inflammatories and pain relief to animals ourselves, without having to get a vet involved," says Robert. "It not only saves time and is much better for the animal, it's amazing how something as relatively simple as a quick dose of Metacam can take the stress out of an illness. Sometimes the trauma of childbirth can be a real shock to an animal, and if you can ease that stress, it's surprising how much better the animal can do."

On occasion, it seems nothing short of a miracle that a newborn lamb pulls through. When some of the tiny tots are delivered, their futures look very doubtful indeed. Take Tiny Lamb, for example; he was one of the smallest lambs to be born at the farm when he came into the world during spring 2020. He was just a quarter of the size of his sibling, and it was really touch-and-go whether he would survive. But even though he was only a slip of a thing, Robert could see that he was a strong suckler – which is always a sign that a lamb has the instinct to survive, so Robert and the team decided the lamb should be given extra-special care. With regular feeds and tender loving care from the farmers, Tiny Lamb pulled through. And, in just four months, he'd grown to be half the size of his fully grown mum.

The pure-white youngster continued to thrive and became such a little hero at the farm that the farmers were pleased that they'd decided he should be a teaser tup – a ram used to woo the ewes into season. "The same goes for Nemo, another one of our lambs that needed extra-special care that we wanted to keep on the farm." Normally, lambs' testicles and tails are docked when they are just a few days old, so the farmers make a call about whether a lamb will be a teaser tup more or less from day one.

In 2021, Tiny lamb was given a vasectomy, which sounds drastic, but it was a simple, painless operation carried out by Shona Searson. "She gave him an anaesthetic so he was pain-free and sat on his bottom while David and I supported him either side," says Robert. "The operation just snips away the tube that carries sperm into the testicles. So he still looks like a tup, smells like a tup and behaves like a tup, but he can't get any ewes pregnant. He won't know he's not 'actually' mating them either because it will feel the same for him, plus he will still have that urge to mate, but his genetics aren't good enough for him to be a breeding ram himself. As a teaser tup, he'll mate with the ewes, then the purebred ram will be put in with the ewes a week later and he'll make them pregnant."

In the past, the brothers have joked that a teaser tup's job is "all of the intent but none of the responsibility". "He's got the best of all worlds as he gets to mate every ewe on the farm, whatever breed he is. So he is the farm stud!" Robert says. Not only that, teaser tups get to stick around on the farm forever, while other lambs may be destined for the farm shop. "We're lucky that if an animal makes an impression on us, we can indulge ourselves to a degree and spare it the fate of some of the others, whereas a strictly commercial farm doesn't always have that option. We're a farm and we understand what farming is all about – it's a business, after all – but, on occasion, we can bend the rules for special ones."

To keep the bloodlines strong in the various breeds of sheep on the farm, new rams are regularly introduced, and the better the pedigree, the more purebred the new lambs will be. So for each variety of sheep, there will be a ram to service the ewes. Arnie for the Jacobs, Bear for the Swiss Valais, Vincent for the Dutch Spotted, and so on. They live together as a bachelor pack, spending summer in a field together and hanging out in the rare breeds barn during the colder months. "We put them all together at the end of tupping," says Robert. "They are all physically drained, so they are much less likely to fight. They haven't the energy to get aggressive and, by putting them together in one pen, it saves the time and space of having to keep them separate. We start off with a relatively small pen, which doesn't have enough room for them to fight even if they wanted to, and they all become pals very quickly. We've found it's a system that really works.

"We are lucky enough to stock rare breeds and we have some animals that are just like family pets, so not all of our animals are bred for the food industry. But, with all of the animals that we farm, we like to think that we are giving them a fabulous life and helping the rare breeds is a bonus. Looking at the bigger picture, having greater numbers of rare breeds in the UK is beneficial for both the education and preservation of farming history, and it celebrates the rare breed's unique attributes."

18

There's nothing like a chorus of enthusiastic bleating to announce a new set of twin or triplet goats has come into the world. And every year in spring, there will be several days when the Cannon Hall farmers will arrive for the start of their working day to find some happy newborns already making their appearance known. Goats often give birth with no assistance needed whatsoever, so once the newborns are safely delivered, their mums have a sniff of them, give them a good lick, and the kids are dry in no time. The farmer then lifts them up to give them the usual spray of iodine on their navel and a quick check-over, then safely pen up the mum and her newborn for them to bond together and for the new kid on the block to have a good feed.

In 2021, it was Philippa the pygmy goat's turn to give birth. Philippa is often referred to as a "gllama" because she prefers the company of the Cannon Hall Farm llamas to her fellow pygmy goats. One Saturday in April, she delivered twins in the barn with her llama friends. "But something a little untoward is going on," said farmer Dale, as he filmed the new family for the farm website. "She usually gets on better with llamas than goats – or maybe not . . ." There was little Philippa looking very pleased with herself and her two very mismatched kids – one was a little Boer goat and the other was a pygmy. "They are definitely not the normal kids that we would see," says Dale. "One very large black-and-white Boer kid, and one tiny cream-coloured pygmy kid."

It seems that, although Philippa usually hangs out with the llamas, she definitely preferred the company of not one but two billy goats at some stage during the back end of 2020, as, around five months later, she delivered two different breeds of goat – just moments apart. She must have been mated by two different billies – one pygmy and one Boer – and, in this case, two of Philippa's eggs were fertilised by different fathers. It's biologically possible for it to happen, but it's somewhat unusual. Mind you, given that goats have a habit of being expert escape artists from wherever they are kept, it's also understandable. Pygmy goats also have a natural curiosity to hang out with other breeds – and, if there's one goat to prove she likes to mix things up with the company she keeps, it's Philippa the gllama.

Philippa's mismatched twins weren't the first of her off-spring, but seemed to remind her that she was, in fact, a goat and not a llama after all. Months later, Robert caught up with her hanging out with the billy goats Goose and Maverick in the fields, so no doubt there would be more kids to come. It's any-one's guess what they will look like . . .

Goose and Maverick were made famous in an episode of *Springtime on the Farm* in 2021. Their story was highlighted as, unusually for billy goats, Goose hadn't been able to perform to his usual standard at the start of the previous mating season. In true *Top Gun* style, his wingman Maverick had been called in to help Goose complete his mission – just without the Ray-Ban aviator sunglasses. In their usual fashion, each of the billies had got themselves ready for romance by urinating all over their bodies so that their pungent perfume would send the females into a fertility frenzy. But it was only when Maverick was introduced to the ladies that Goose sprang back into action and realised that he had been put into the barn with them for a reason. "When I saw Goose strutting his stuff, I can admit, it took my breath away," says David. [Please David, no more *Top Gun* puns . . .]

The following spring, the fruits of Goose and Maverick's labours were plain to see, with a kindergarten of new healthy kids frolicking in the barn. The good news is the two fathers continue to perform well at any given opportunity. Goose obviously got his mojo back when Maverick appeared on the scene. "They are really doing well," says David. "Maverick is our older billy goat out of the two of them, but is still looking young and virile, and doing a fantastic job. The thing is, Goose and Maverick are the fathers of a lot of the goats on the farm, so we're planning a fresh bloodline to produce some new ones. We kept every female kid that was born, so we will get another couple of new billies to mate with them and will keep breeding Goose and Maverick with the older females. We have around seventy pygmy goats on the farm now and we love them. They're so much fun and they're so cute and entertaining – some of the best animals we have."

The goats have also proved to be very profitable as they are sold on to goat breeders and to individuals as pets. After the kids are weaned, they are tagged and photographed and, when the farm gets an enquiry from someone who'd like to buy one of their pygmy or Boer goats, they're sent a selection of pictures to choose from. The individual goat is then allocated to the buyer, which is obviously a lot easier than coming to the farm and trying to select a goat while they are enjoying their adventure playground.

"Goats go mainly to smallholders and as pets," Robert explains "but we don't sell them to people who we don't think are equipped to look after them or don't have the facilities to give them a good home. A goat needs a paddock or at least a big garden that they can run around in, and a shed to shelter in because they don't like getting wet. We've found that if a potential buyer doesn't have the capacity to look after the animal or doesn't have the right set up, it will just end up coming back to us anyway." It's like adopting a dog without considering how much space and time it takes to look after it. "We sell Boer goats too," says Robert, "but they tend to go to commercial goat farmers, so it's more likely that we would sell a batch of ten breeding females. They produce some milk, but are bred mainly for their meat."

"Goats make great pets," says Robert. "You feed them hay in winter and grass in summer and they like a bit of goat muesli, which is dried cereal and molasses – a bit like Country Store for goats. Delicious. What's not to love?"

All farm animals are tagged at Cannon Hall Farm as it's a legal requirement to be able to identify which farm they are from, in a similar way to microchipping a pet. "We can't lie, it's not totally pain-free when we tag an animal," says Robert, "but a skilled operator knows how to minimise the pain by avoiding sensitive areas of the ear. It's like getting your ear pierced. Every goat, sheep, cow, pig, llama and alpaca that we have has to be tagged for tracing. It's not something that we have a choice in doing; it's just farming compliance," he explains. "Every animal has to be identified so that there's a record of them if they move into a different area. Then, if a disease breaks out, they can help trace it back to the farm it's come from." In the case of historical outbreaks of foot-and-mouth and bird flu, it could be quickly established where the disease originated, because animals were tagged, meaning the outbreak could be contained much faster.

—

Because the laws surrounding farming change fairly frequently, farmers continually evolve their working practices and the methods for the way they work with different kinds of animals. At Cannon Hall Farm, just because things have worked one way for a number of years doesn't mean they won't try something new. By adapting and varying the ways that the farmers operate, they are continually learning new things about their animals at Cannon Hall Farm and sometimes the team surprise themselves with how well things work out . . .

In April 2021, Helen the alpaca was finally introduced back to the family group of other alpacas. She was hand-reared by the farmers as she'd had a shaky start and wouldn't feed from her mum Audrey. She had a long spell in the special care unit and for a lengthy spell afterwards, she lived in one of the rare breeds barns with pygmy goats Mille and Primrose. When Helen was first introduced back into the alpaca family group, she met with some resistance from feisty female Shakira. You couldn't help but feel for Helen – she's such a docile friendly softie and Shakira wasn't exactly welcoming. But it seems that Helen's laid-back personality started to rub off on the other alpacas. The group went from being stand-offish and naturally territorial to positively welcoming to the farmers when they went into the alpacas' pen. The farmers barely had to worry about being showered in foul-smelling spittle. Waterproofs are always at the ready, though; there's no point in taking any chances . . .

"Whenever we introduce a new animal, we always have a long discussion with the breeder that we are buying it from about the correct way to help the animal flourish on the farm," says Robert. "Many of our farmers have studied animal care and know the specifics of what each breed needs, including its diet, habitat, healthcare and behavioural needs, so we fully trust that they are in excellent hands." As to more specific knowledge – things like how you restrain a llama when it's showering you in spit or calming down Robert's nemesis, Zander the alpaca – you just learn as you go along. "Google is incredibly useful and now you don't have to be ignorant of anything as you can check things as needed," says Robert. "But, generally, if you have an understanding of farm animals, I think your skills are pretty transferable. It's about having an innate feeling for animals, and our best people are the ones who are really interested in the breeds they look after."

The Nicholsons have also learned how to professionalise their breeds. When the farm first opened, it was just a case of buying and borrowing as many different kinds of animal as the family could lay their hands on, so that they had a good number of animals for visitors to see. Now the breeds in their collection are much more considered. When it comes to expanding the numbers, there's much more research done before a purchase is made.

The farmers also constantly try to keep abreast of new developments in farming. "Originally we learned everything from my dad but he's always learning too." Robert says. "For example, back in the old days, farmers didn't use stomach tubes to feed vulnerable newborns, but we do that all the time now so that milk can quickly be deposited where it needs to be."

Wet adoption is another practice which has become part of the usual lambing tried-and-tested developments, as is rehydrating animals that are having difficulty birthing. It's a matter of inserting a tube into the uterus and dispensing a mixture of warm water and lubricant. As Robert explains, "If a ewe is short on moisture and it's very uncomfortable and difficult for them to get a lamb out, this is a kind way of helping her along. In the case of an animal having to birth a stillborn, it's the most sympathetic way of dealing with a sad situation.

"One of the most useful things that we have learned about is why sheep foot trimming isn't always necessary. I watched a farming programme and a vet said that if a sheep is infected with foot rot and the foot is trimmed, it just causes the infection to spread. The vet's advice was to treat the sheep with antibiotics, isolate the infected sheep from the rest of the flock and slowly reintroduce it back into the flock. We started doing this about eight years ago and, sure enough, our foot health on the farm has never been better. Dad initially wasn't too sure, but he's coming round to the idea!"

19

When there's non-stop lambing activity in the Roundhouse, plus all of the other planned, and not quite so planned spring births, the woodland around the Nicholsons' farm provides a welcome haven of calm. During the 2020 pandemic, the need to able to to take a deep breath and recalibrate away from the constant drama meant the wood became even more important than ever. A happy mindful place for whoever might need it.

Much of the woodland was felled and replanted in the 1980s by Richard, Robert and David, who cleared a quantity of fir trees and planted native English oaks and lime trees. "We also repaired the pond, which had burst its banks, packing clay around the edges of it to save it from collapsing in on itself." As kids, the pond would have been the first place where they tried their hands at fishing and was always their favourite place to find frogspawn in the spring.

After the closure of many coal pits in the area following World War II, large swathes of the land around Cannon Hall Farm were chewed up by diggers in a practice called open-cast mining. This was a harsh technique to extract minerals near the surface of the land, rather than mining deep underground. It may once have been considered necessary, but the landscape looked very different as huge machines tore into the earth, turning what were once unspoilt fields as far as the eye could see into an eyesore. Even areas of the beautiful gardens in the park at Cannon Hall were destroyed.

While coal mines had established countryside communities and provided a living for thousands of people, from the word go, open-cast mining wasn't popular – therefore, from the 1960s onwards, the new method for extracting minerals from the land was on the decline. Landscapers were employed to repair and replant destroyed areas and restore the countryside's natural beauty. One such contractor was the brothers' cousin Richard, who worked on the restoration projects in South Yorkshire. In 1986, when David left school, he got a job assisting him and became quite adept at tree planting.

This came in handy when the family decided to work on their favourite area of woodland. Once Roger had decided which trees he would like to be planted, the brothers went along, armed with the saplings. "We went for English oak and lime trees, which are two of my dad's favourites," says David. "Once you find the right spot, it's just a matter of digging your spade in, easing it forward, sticking the sapling in the space, patting it down, and you're away. That's a tree planted. There's nothing difficult about it and that tree will be there for years and years to come."

With the help of the natural environment and the wildlife that lives amongst its leaves and branches, woodland generally maintains itself, but every so often, it needs a bit of a helping hand. When the trees thicken in warmer months, the forest becomes overgrown, and the lack of sunlight causes the land to become boggy, which is less attractive for its normal residents of insects, birds and small mammals – all key links in the food chain. Therefore, frequent judicious pruning is regularly carried out to help the woodland flourish. This is generally done early in the year when trees have less foliage, so birds will not yet have made their spring nests and won't be disturbed.

In April 2021, Robert and David got ready for a morning of tree surgery, cutting back some of the weaker trees to give the stronger ones better access to sunlight and increase their share of nutrients from the earth. With David in charge of the chainsaw and Robert in full voice to shout "Timberrrrr!" the brothers got to work. If you watched the tree-felling episode of *This Week on the Farm*, you will have seen how horse logger Andrew and his trusty Dales pony Charlie cleared the timber from the forest floor. With the trees densely packed together, it wouldn't have been possible to get a vehicle into the forest to load it up and take it away without destroying vegetation. So the strength, dexterity and horsepower of Charlie the pony was vital to carry out the task.

But here's a little secret: a different pony was initially used when Robert and David went to film the woodland story. "The first pony started running amok!" says Robert. "And one of the production team went flying as the horse was a bit spooked by the camera filming him. It was bedlam, a real 'oh 'eck' moment." Luckily, the team was able to call up a replacement horse logger and Charlie the pony came to the rescue. With the new wonder horse in place, filming was able to continue.

"Even though it was a bit dramatic at the time, I was glad that we could make the story, because it was lovely to be able to highlight the work of Dales ponies. Logging ponies really are the archetypal work horses as they used to work underground in the mines and above ground on the farm. Horse logging dates back over ten thousand years and really fits into the rhythm of country life, so seeing Charlie walking up through the woods really lifted my heart."

For good measure, Robert and David planted new yew saplings to help provide more warmth in the lower levels of the woodland, which in turn provides shelter for mammals such as badgers and hedgehogs. "Planting trees is something we have always been passionate about," says Robert. "Like the saying goes, the best time to plant a tree is a hundred years ago and the next best time is tomorrow."

As a family of tree lovers, the Nicholsons share a particular love for the stunning oak tree that's the jewel in the crown of the paddock at the bottom of the farm. The brothers will even argue amongst themselves whose favourite it was first (Richard wins as he's the eldest!). In spring 2021, Roger decided to continue the legacy of their favourite tree and collected some of its acorns, planted them in his favourite compost and grew them on in his greenhouse. Lo and behold, each one soon sprouted into a strong sapling, which will be replanted around the farm.

"In two hundred years, if we've ever got any generations of our family left, they'll look at them and think, 'Someone must have really cared,'" says Roger. "I'm hoping that there will still be technology around for them to see what their great great-great-great-great-grandparents did here on the farm. It would be wonderful to think of them enjoying it as much as we all do now."

—

As every Cannon Hall Farm fan knows, the Nicholsons' land is home to a myriad of animals and the family is always happy to welcome a new breed. Covid-19 scuppered the chance of introducing some of the new species that the team had hoped to home at the farm in 2020, but, after 18 months of planning and paperwork going back and forth, in April 2021 farmer Ruth set out on an exciting road trip to pick up some new exotic guests. Exciting as a trip up the A64 can ever be . . .

Although the banded mongoose is native to South Africa, Ruth only had to go as far as Flamingo Land Zoo in North Yorkshire to fetch the four new furry friends destined for Cannon Hall Farm. After a quick chat with the keepers and an intel session about any individual issues to look out for, the mongooses were soon loaded up and on their way to Barnsley. Unluckily for Ruth, a pungent, windy smell continued to circulate around the car for the entire journey home.

Back at the farm, in a film for the Cannon Hall Farm website, Ruth pointed out that the new mongooses have similar needs to the meerkats, just with slightly sharper, pointy teeth. So they may look cute, but a bite from one of them isn't worth any amount of cuddles. The team knew that they would have to tread carefully and let the four male mongooses settle in nice and slowly. In readiness for their arrival, a fantastic new habitat complete with underground tunnels had been created for them. And very quickly, with a few tasty treats, they were soon making themselves at home. It wasn't long until the four curious friends were sussing out their new surroundings and marking their territory.

Like meerkats, mongooses live in family groups, which can range in number from around seven to forty individuals. But unlike meerkats, who usually have one dominant male and one dominant female in their group, all of the female mongooses in the pack will breed – sometimes with several males. In the wild, they live in abandoned termite mounds but, if they are unable to find a suitable abode, they will pile on top of one another to sleep, with the larger adults facing outwards, checking for predators. Once the mongoose gang got established at the farm, it was hoped that they will follow suit and pile up together as it was unlikely they'd find an abandoned termite mound in Barnsley.

Within weeks, the new mongooses were happily chirruping

away as farmer Kate arrived to feed them their breakfast of crickets, courgette and cat biscuits. "They're not friendly animals," says Kate. "But they're happy for me to be in their environment because I'm feeding them. I'm thrilled that they are here as it was such an empty enclosure for so long because we weren't able to go and collect them. The fact that we were allowed to go and pick them up meant that we were properly getting back to normal again."

Covid-19 put the brakes on several projects that were planned at Cannon Hall Farm. But, with no visitors from spring 2020, it was also an opportunity to work on new and improved facilities – for both humans and animals. While the brand-new Lucky Pup restaurant was being completed at the entrance to the farm, a new ferret enclosure for the racing males was also being created with the help of Cannon Hall Farm's incredible team of carpenters and builders. "In as much as we can, we will always try and recreate our animals' natural environment," says David, "as well as taking into account our rigorous health and safety rules on the farm. We want our visitors to be able to enjoy the animals and get as close up to them as they can without being in any danger. Some of our animals' teeth are very sharp – as I've discovered to my cost."

In April 2021, it was time for the ferrets to move into their luxurious new home. It was positioned closer to the ferret racing arena, meaning that once Covid-19 restrictions were lifted, visitors to the farm would once again be able to watch the ferrets going head-to-head in the tube tunnel race. With a huge new pen for them to run around and play in, a purpose-built warm underground area for them to sleep in and an alfresco dining area, it was a grand design for the nine farm favourites. "We decided to make them work a bit harder for their food," says Kate. "We built a ramp up to a bird-house type structure where their food is. We've designed it that way to encourage them to keep running around and being fit and healthy. Their enclosure is the first thing that visitors will see when they come into the farmyard and the ferrets are always very entertaining. So everyone will be happy!"

At Cannon Hall Farm, the farmers don't need to be too concerned about predators on the farm as the small mammals are very good at looking after themselves. Ferrets, mongooses and meerkats all like to snuggle up and sleep together, and, if a fox ever did find a way to get to their enclosures, they would be likely to get short shrift from the small mammals with their territorial pack personalities and those savage sharp teeth.

Meanwhile, out in the fields there are other deterrents. "Magpies and crows can be nasty opportunists and can peck at a vulnerable sheep, like the one that attacked the ewe with a prolapse at Mill Farm," says Robert. "But we've found a very successful way to keep predators at bay. The llamas like to be in the fields as soon as it gets warm enough for them to enjoy it and, if we need a guard animal, Elvis is an absolute legend. If we were to have any problems – and we seldom do – then Elvis would soon see it off with a stampede and a deluge of vomit-smelling spit. Also, having llamas out in the fields means they'll eat away at the coarse hedges and it wears their teeth down nicely, so it's a win–win for us."

20

Spring has truly sprung when you see a swallow. With their perfect aerodynamic shape and their unmistakable long tail streamers, they're a true sight to behold as they announce their return from Africa. The old saying is "One swallow does not make a summer" but it certainly makes a spring. "I'm always excited by the sight of the first swallow," says Cynthia. "We used to get a lot more than we do now and they would all be lining up on the telegraph wires next to the house."

Reaching speeds of up to 65 kph, a swallow eats on the wing, consuming around 60 insects per hour or a whopping 850 of them every day. It's no wonder they need to ensure it's warm enough for them to be able to return to their favourite nesting places. Without the flowers being in bloom, there won't be the insects to feed off them, and without the insects, no swallows. But return they do, year after year, often to the same colony and often to occupy exactly the same nest that they were born in. What makes this particularly remarkable is that they will have flown around 10,000 miles and they still have enough energy reserves to rear a clutch of between four and five eggs. They are the Duracell bunnies of the avian world.

Since 1970, swallow numbers have been in decline, due to climate change and changes in farming practices. With plants coming up out of the ground earlier, but fewer insects to propagate them, nature is facing many new challenges. Nevertheless, in spring 2021, there was still believed to be a healthy population of 1.4 million pairs of swallows that return annually to the UK.

Knowing that swallows typically visited the farm each year, but with no evidence of their nests in the farmyard, in spring 2021, Robert and David fitted new purpose-built swallow nests right in the heart of their land, underneath the roof of the pig shed. The brothers were joined by wildlife garden designer Joel Ashton, who showed them how to use modelling clay and straw to replicate the mud and vegetation creations that the swallows themselves make. As usual, the competitive duo tried to out-do each other with their building skills. And once the new nests were in situ, Robert and David not only challenged each other as to who would see the first swallow of spring, but whose nest will be occupied first.

In the end, neither of their handmade homes made the grade for the discerning swallows at Cannon Hall Farm. The feathered friends came back as normal, but favoured their old haunts in Roger and Cynthia's garden. Maybe it was the abundance of flowers in the garden – and therefore plenty of insects – which tempted them back?

"We don't really give it much thought, to be honest, but I suppose we're very lucky in the birds that we have here," says Cynthia. "We get swallows, swifts, house martins, chaffinches, dunnocks, robins of course, wrens, nuthatch and blackbirds aplenty. And we get pied wagtails, which are always a delight. We used to have a cat called Jimmy, but he kept killing the birds, which I never thought was very fair as we did feed him – we didn't leave him to fend for himself. He kept down the vermin, though, so I suppose I'll forgive him."

—

As if we could all forget, the second national lockdown continued into spring 2021, and once again people turned to nature to find respite from the continuing bad news. Robert re-enacted the routine he'd perfected in the first lockdown: 3am alarm, walk around the farm, bacon sandwich, pint of tea, hot bath and a daily Facebook broadcast. Meanwhile, the rest of the farm was its usual hive of activity – albeit with no visitors to enjoy the animal shenanigans. And, once again, some staff members switched jobs from working in the restaurants to helping with online orders in the farm shop.

"I carried on as I had in 2020 and it brought me stability," says Robert. "It became really important to me so that I could collect my thoughts and plan my day. Saying that, though, the second lockdown certainly felt very different to the first one. We felt much more organised in the sense that we had done it before so we were used to the mask wearing, the social distancing and all the preventative measures we had to take. It was a nuisance, but the vaccine was there so at least we had hope. Our main focus was getting Mum and Dad vaccinated and getting them both the hospital care that they needed as they both had other health issues."

With Roger out of action, the brothers had to step up to the plate more so than ever before. And there was no let-up with new series of *This Week on the Farm* and *Springtime on the Farm* in the offing, the publication of their first book, as well as the usual hatches and dispatches of spring. "Usually we have Dad doing his Ranger Roger bit and spotting any potential problems with the animals, but in 2021 it was up to us, which we were all a bit nervous about. But the gods must have been with us because we've never had as few cases of mastitis, coccidiosis and orf. We considered ourselves extremely lucky because outbreaks can spread so quickly. It could be that, because of lockdown, we had a bit more time to spend on the animals and the weather certainly helped. It was kind and dry. If it hadn't have been such good weather, we could have had a right job on our hands."

When any of animals died during lockdown, it felt doubly painful to deliver the news via the farm website or on Facebook broadcasts. "The more that people become invested in an animal, the more difficult it is when we lose them," says David. "We almost feel like we're letting our followers down if one of the farm favourites dies, so we try even harder with them. It's always difficult when you lose any of our animals that have become like pets to us. It's also great to be able to make people happy, so we really do try our utmost."

—

There was happy news from the calving barn in April 2021 when prize Highland cow Fern gave birth to a beautiful female calf, who was later to be named Lilibet. "It was the most adorable little calf and we were all really delighted," says David, "but unfortunately the calf was ten days premature and wouldn't start drinking from its mum, which meant there was a lot at stake. If it didn't start getting its vital colostrum, we could have lost it. We had to act fast."

Knowing that a normally placid Highland cow can turn into a very protective mother, the brothers had to tread carefully to be able to help the new calf. Robert wasn't taking any chances.

"There are two areas of concern for me," he said, as he reported the news on the farm website. "Her horns are big, they're strong and she knows how to use them. Plus she can kick like a donkey with her back legs. But thankfully for me, I'm involved in neither of those areas. This time, I'm not going to get that close to either end."

Even with Fern safely stored in the cattle crush and all manner of gentle coaxing to get the calf to feed, it still wasn't happening, so farmer Wade carefully milked Fern, capturing her vital colostrum. Letting her have a few hours' rest, it was hoped that later Fern would be more receptive to allowing the calf to feed from her. Robert was determined that they shouldn't give up too quickly and start hand-feeding the calf. "Once a calf gets used to feeding from a bottle, it makes suckling from the mum more difficult, but hopefully it will smell the milk on mum's teats, know what it is and go back to her."

A week later, the calf was suckling, but not on all four teats, which led the farmers to think that Fern could possibly have an inflammation that could lead to an infection, such as mastitis. David had to manhandle Fern back into the cattle crush to encourage her to feed her calf from that particular "quarter" of her udder [there are four teats on an udder], so that it would keep the whole udder in tip-top condition. "We had to convince her that it's in her best interests to allow the calf to feed from that teat," Robert explains. "With it drinking, it will reduce the inflammation, make it less tender, and make it a pleasurable experience, so the calf gets its full quota of milk."

And, after a fair bit of cajoling, dedication and care, the little calf finally got the hang of suckling from every quarter. The farmers could breathe a sigh of relief. "The calf needed to be carefully fed and watched and its future was by no means guaranteed," says Robert. "So, to see it drinking, thriving and growing every day is a major boost for the farm."

It's not uncommon for a mighty animal like Fern the Highland cow to be protective of its newborns, but bulls have the worst reputation when it comes to being aggressive. Interestingly, though, it's the often slimmer and smaller dairy bulls that should always be avoided. "They can look very pretty," says Robert, "but Jersey bulls are the worst of the lot and can be really nasty. Legally, you are not allowed to put a dairy bull on a public footpath, although a beef bull is allowed." So, while they may look formidable, that bull that you encounter on your country walk may actually be as gentle as a pussycat. That said, maybe it's best not to offer it a saucer of milk. For more information on the Countryside Code, see page 267.

21

At the end of April 2021, the Swiss Valais twins Brussel and Sprout also got their first taste of spring grass. But getting them into the field wasn't without its difficulties. Even though there were just two of them travelling there – Club Class by any other animals' standards – the dynamic duo were reticent to board the ramp up to the trailer. "Swiss Valais Blacknose sheep are super friendly," says Robert, "so they're harder to load than ordinary sheep because other sheep naturally have more fear and will run away from you. We literally only moved Brussel and Sprout a hundred yards from the Roundhouse into the centre of the sheep-racing circuit. But what a prime move it is, into the sunshine. There's some lovely grass for them, and they are going to have a fabulous time there."

Turning out the new lambs on to the fields for the first time is a joy for animals and farmers alike. Once it's warm enough for them to have enough grass to graze on, and there's no more hideous weather forecasted, then it's time for wagons to roll.

But, first, the lambs have to be loaded up – and that's where the clever design of the Roundhouse really comes into its own. Sheep can be corralled from one pen into the centre of the Roundhouse, then ushered through the back of the structure and up a gangplank, straight on to the trailer. Once it's full, the sheep are secured inside the trailer and off it goes. When the farmers are transporting family groups, to ensure that they end up together when they arrive at the field, they carefully carry the lambs up to one section of the trailer and their mums travel in the back section. "We contain the lambs separately from the ewes so that they can't get trodden on as the ewes stampede into the trailer," David explains, "then we run the ewes up to the back of the trailer. The ewes and lambs are all marked up with corresponding numbers or letters that show which family group they are part of, so if there's a problem in the field and one of them gets poorly, we can bring the whole family back in together."

It's a noisy process as the sheep are often panicked by the sudden change of scenery, the change of air and the rumbling sound of the farm vehicle. "The fact that they make such a noise is actually a really good sign," says Robert, "as it means that the ewes and lambs are really missing each other. When they get out in the field, we need them to find each other, spend time together and mother up properly and get that family unit back together. It's a hop, a skip and a jump to us, but to these sheep, it's a whole new part of life's journey."

Once they reach the field, the lambs are carefully lifted out of the trailer first before the ewes are allowed to join them. "The last thing we want is for the ewes to run out and get separated from their lambs," says Robert. "The ewes should find them straight away and they'll re-establish the bond." The ewes delightfully trot down the ramp and on to the springy grass and it only takes moments before the mother and child reunions are in full swing. "They find each other by the sound – they know each individual cry and each other's smell and when it all goes quiet you know they have reconnected. They will then wander off, nibble some grass and find some friends to play with."

It's a magical sight to see the young lambs taking to their new surroundings so quickly. It's almost as if there are actual springs in the spring grass as the beautiful lambs bounce and bleat in delight. But who wouldn't be happy to have acres of space to run around in, top-quality food at their disposal and fresh, clean Yorkshire air?

While the newborns are having their first taste of life out on the grass, over on an adjacent field, the February lambs are ready to be vaccinated. In the same way that their mums were vaccinated in January, the two-month-olds are given an injection of Heptavac to shield them against the seven main diseases in sheep. All of the lambs are also wormed for the first time, and will subsequently be wormed again, around four or five times a year."

All grazing animals are susceptible to worms, so the farmers test their faeces every six or so weeks. It's collected up and sent to the vets and, depending on whether there's a build-up of worm eggs, they'll be treated as necessary.

Vaccinating and worming is heavy, physical work as each lamb has to be lifted up individually for the worming solution to be squirted directly into their mouths and the Heptavac to be injected. Because of the sheer numbers to be treated, the lambs are penned off in batches before being allowed back into the field to rejoin their mums. It's a job that takes all day and will have to be repeated six weeks later to ensure that the lambs are 100 per cent vaccinated from disease. "The job is never finished when you're farming," says David. "There's always plenty to do."

Then there's the dagging to get done, which is trimming off areas of the coat covered in muck. "In late spring, it's warming up; flies can lay their eggs on the poo and maggots can infest the sheep," says David, "so, by cleaning up their mucky bottoms and snipping away daggy wool, you take the muck away as well as the problem." With the sheep looking trimmed and pretty – and more importantly free of potential pests – they can scamper off to rejoin their friends and family.

—

However strong the bond is between mother and young, and however well the lamb, calf, kid, cria, pup or other newborn has been able to suckle from its mum, there comes the inevitable time when the youngster needs to be weaned. Its nutritional needs change as it grows, so a mother's milk is no longer enough to sustain it. It's vital that they can survive independently and learn to feed from other sources. The animals that live at Cannon Hall farm are perhaps luckier than some as they often stay in family groups; however, it's crucial that each of the youngest generation is correctly weaned, for a number of reasons . . .

In some cases it's so that the mother can return to their optimum health in order that she can produce more young. In the case of Orchid's foal Blossom, she was moved away from her mum because Orchid was not letting her get close to her to feed. It's a natural behavioural pattern to occur because Blossom was old enough not to rely on her mother's milk and she'd been distancing herself from her mum anyway. Farmer Ruth decided it was the time to separate mother and daughter properly for the weaning to take place and, although it was stressful for them at the time and they called out to each other in distress, they quickly adapted to life without each other. And once Blossom was weaned on to solid food, she was able to rejoin Orchid and the other Shire mares out in the paddock.

Different animals are weaned at different ages, so it's not necessarily a springtime task. February lambs will be taken to graze on grass out in the fields and their milk will be supplemented with solid food known as creep. Creep is emptied into lamb feeders, which are specially designed to be too small for their mothers to feed from, but perfect for little lambs. "Even though there are so many lambs and sheep in a field together, sheep will remember their mums until they are weaned," Robert explains. "But once they have been weaned, no animal will necessarily remember who their mum is. Horses seem to remember the most, but, when they are reunited, it isn't a case of them throwing their hooves around each other; there may just be a whinny or two to say hello."

"The weaning process isn't without its stresses," says David. "It's better if a young animal is taken out of earshot from its mum so that they can't call to each other, as most animals will be upset when you separate them. But if you left them together too long, they would naturally grow apart anyway, and there would be a point where the mother would drive them away – which can be upsetting for their offspring." Young males need to be weaned from their mothers as they can ultimately become quite aggressive. They also need to be kept separate from other females as they will try to mate with them.

"In the case of piglets, they'll stay with the sows for about five weeks," says Roger. "After that, they are moved to a pen away from Mum. By this time, the mum is usually quite glad to get rid of them, because they grow so quickly. If you've got fifteen piglets piling on top of you, any mum is likely to get a bit fed up and ratty about it!"

22

At the end of April, the Roundhouse is a much calmer place to the hive of activity it has been. All around, new mums are bonding with their lambs. Some of the more vulnerable are being hand-fed by farmer Kate or happily latching on to the automatic milking machine. For once, there's time to actually take it all in and enjoy the rural scene rather than rushing around, making sure that nothing untoward is happening. But, just as there will have been some premature lambs in the season, there will be some late ones too.

In 2021, the farmers were patting themselves on the back for having one of the most straightforward lambing seasons that they could remember. They'd had their fair share of seemingly never-ending nights, breech births, the odd stillborn delivery and other complications, but overall, it was a textbook spring.

"It's inevitable, with the volume of sheep we have on the farm, that we will have some stillborn lambs," Robert explains. "Sometimes in a natural birth, the lamb can be born in the fluid sac, and, if you are not there to save it, it will breathe in fluids and die. Disease can also play a part so we routinely vaccinate ewes for things like toxoplasmosis, which can cause ewes to lose their lambs. Even when we are doing the actual vaccinations that will help prevent diseases, the ewe can get stressed and run off, and you don't want them bumping into anything as that can cause a lamb to die. Basically, you have to wrap up your ewes in cotton wool in late pregnancy."

Inevitably every spring, there will be a final ewe to give birth, and, in 2021, it just so happened that it crossed over the time of day when Robert had been on lambing duty and when David was about to start his shift. "Rob called me saying that the last ewe was looking like it was going to need a caesarean," says David. "Apparently she'd been trying to lamb for three hours and Rob had done all that he could to assist the birth, but it was definitely time to get the ewe to the vets. He told me that the ewe was in a pen by herself and I loaded her into the trailer without a second thought and off I went. I'd only got about half a mile down the road when I got a very confused call from Ruth. I told her that I was taking the ewe to have a caesarean and she said, 'I was just about to inject her because she lambed yesterday! You've taken the wrong one!' No wonder the ewe had looked a bit confused, but, as usual, I'd been working at a hundred miles an hour and was determined to get her to the vet quickly. I would have felt a right idiot if I'd got her all the way and she was empty!"

"We were that busy and we got our wires crossed," says Robert. "I was referring to one sheep and David was thinking of another, but thankfully we spotted it in time and all was well."

Having located the right ewe – and to be fair, they do all look pretty similar – off David went on his rescue mission. "Sure enough, the lamb needed to come out of the side door, so the vet David Melleney got to work," he says. "It took a while to get the lamb out and he wrapped it in a towel and passed it to me. It was pitiful to see because the lamb threw its head back and it was thrashing around because it had obviously ingested fluid. I worked on it for about two minutes and tried everything I could think of and all of a sudden its head came up. It blinked and bleated, and I thought, 'Fantastic! It's alive!' But then it crashed again. I managed to get it breathing again, but a moment later its heart stopped again. All of the fluid drained out of its body and I knew it was gone for good. It was so disappointing because my heart was in my mouth while I was trying to revive it and I did everything I possibly could to save it, but it just wasn't meant to be."

There is a happy end to the story, though. David the vet reached back inside the ewe and there was another lovely healthy lamb waiting to come out, so he carefully delivered that one and concentrated on saving both the new lamb and the ewe, then stitched her back up. "It was ironic," says David, "as Robert and I had just been talking about how lucky we were not to need any caesareans for the whole of the lambing season, then the very last one to lamb needed one."

Mum and baby were soon well enough to go back to the farm, and within weeks they were gambolling about, enjoying the fresh new grass in the White Mare field. "It's so rewarding when it goes really well, but so disappointing when you lose a lamb," says David. "But I couldn't have tried any harder. It just felt particularly significant as it was the very last ewe of the season to lamb."

—

Even experienced farmers like Robert and David sometimes need help with lambing issues, so they were only too happy to help with one of the UK's youngest farmers, Joe Trofer-Cook, or Little Joe as he has come to be known. In 2020, Joe's parents got in touch with Cannon Hall Farm saying their eight-year-old son, Joe, was a big fan of the Nicholson brothers and hoped to one day follow in their farming footsteps. During the first lockdown, Joe started selling his home-grown vegetables from a wheelbarrow at the bottom of the family garden to make money to buy food for his chickens and sheep. His determination and hard work so impressed Robert and David that they decided to surprise him at his parents' farm in Lincolnshire.

Little Joe's face lit up when he saw Robert and David suddenly appear while he was feeding his sheep and he was absolutely speechless with what happened next. He'd written to Santa Claus asking for a cow for Christmas and his parents had got together with Robert and David to make his dream come true. The little lad couldn't believe his luck when Rob and Dave opened the back of their trailer to reveal a beautiful pair of crossbred Aberdeen Angus calves. And, even though he'd had no experience farming cows before, once Robert and David showed Joe how to handle them, it was obvious he was a natural.

The following spring, Joe visited Cannon Hall Farm as three of his Jacob ewes were pregnant. Knowing that Robert and David were better than any farming textbooks, he wanted to get lots of tips for his first experience of lambing, and the brothers were only too happy to oblige. Joe watched them closely as they assisted the delivery of a newborn calf, carefully taking in all the techniques he would soon have to master with his own ewes. As Robert gently lifted a newborn away from its mum, David said, "Clean around its head when it's born and make sure there's no sticky liquid," then Joe rolled up his sleeves to get close to a newborn. He was completely fascinated as the ewe and lamb started bonding but couldn't help questioning why one of the lambs was covered in yellow birthing fluids. "It looks like an egg yolk's been splattered all over it," he said.

A few weeks later, with the imminent arrival of Joe's first

lambs, Robert and David travelled across the country to Lincolnshire to help out their favourite young farmer. In the final episode of the 2021 series of *Springtime on the Farm*, little Joe described how amazing his first experience of lambing had been. "I'd love to send Rob and Dave a picture of Rhubarb the ewe to show how well her lambs are doing." But there was no need for a postage stamp as the brothers were there to surprise him in the flesh and together they watched Joe's new spring lambs enjoy their first taste of fresh grass. The lambs couldn't have looked happier – and neither could little Joe. "It's so unusual to see someone of his age with such a strong work ethic," says Robert. "But I was just the same at his age. At springtime I couldn't wait to get out of school and be out lambing the sheep and looking after the stock and learning how to be a farmer. Little Joe is so old-fashioned in his outlook and so hardworking that he's now a local legend. Everyone wants to support him and see that he does well."

While they were visiting, Robert and David bought some plants from Little Joe for Roger and Cynthia to grow in their greenhouse, including tomatoes, aubergines, peppers, squash, sweetcorn, marrows and cucumbers. And Roger was delighted with his healthy new crop. "He's a clever little lad, is Joe," says Roger, "a proper farmer in the making. But he's going to have to find a way to enjoy school a bit more because he's there for a long time yet!"

Roger was just 16 years old when he started working full-time at Cannon Hall Farm. Tragically, his father Charlie died one year after buying the new farm. "That time is a bit of a blur," says Roger, "but I do remember when I started out as a farmer I made quite a few mistakes! But that's all part of farming really, you learn as you go along. I so admire Little Joe for his ambition and determination. The whole generation of young farmers are the future of farming and they need as much help and encouragement as we can give them. It can be a very tough life, and if you're not one hundred per cent committed, it's probably not for you."

It was back in April 1958 that Roger's father Charlie Nicholson first bought Cannon Hall Farm. So each April is a particularly significant month for Roger. The day that the hammer came down at the auction for the farm had been a day of anxiety, celebration and doubt. Anxiety that Charlie would be able to afford the farm and ultimately finding himself paying over the odds for the property; celebration that his family had such a wonderful new home and business, but doubt that he would be able to make a success of it.

"Even though the farm isn't big by today's standards, it was quite an undertaking back then," says Roger. "We had to go from a standing start in terms of putting crops in and rearing animals, and it took us nearly a year until the farm was ready to move into. Sixty-plus years on and it's still a work in progress."

PART FIVE
May

23

On a wing and a prayer and a dose of good luck, spring will have reached its zenith by May. Gardens will be in bloom, the sound of lawnmowers will be part of the soundtrack to every weekend all over the UK, and, in South Yorkshire, it's even warm enough to shed a few layers – or to at least unbutton your cardigan. At Cannon Hall Farm, with all of the lambing done and dusted, it's time to turn to the next lot of tasks in hand, as always consulting the weather forecast first to make optimum use of the sunny days.

It used to be that 1 April heralded the start of lambing time and on 1 May all of the farm cattle would be turned out to the fields for their spring and summer holiday. However, with weather conditions generally being warmer earlier in the year than they used to be, at Cannon Hall Farm, the cattle are already likely to be enjoying the new grass by May. "It's wonderful seeing the cows with their calves charging into the fields," says Robert. "They really love it when their feet touch the turf for the first time. Any animal loves a wide open field to explore and the very first time they go outside is a big excitement for it."

With a lovely lush paddock right opposite the second rare breeds barn, there will be days out for all of the animals who like getting a good run around. When his hooves touched the soil for the first time in spring 2021, young Shetland pony Hadley took to it as if he was competing in the Grand National, whinnying with delight and charging around like a jumping jack firework. Meanwhile, after an initial burst of energy, his father, Jon Bon Pony, just chomped on the spring grass and looked at the youngster as if to say, "Calm down, dear. I'm trying to have a nice quiet meal here."

Each year, when Shire horses Blossom, Lottie and Ruby get into the paddock for their first day on the new May grass, it's pure poetry as they canter up and down, taking in their new surroundings. "By this time in spring, our Shires will have been in the sand paddocks, so they're already in the fresh air," says Robert. "But there's nothing like the feeling of grass under their feet."

Farmers Ruth, Robert and David carefully lead the Shires into the paddock, making sure to leave enough space between each of the horses in case they kick out in excitement and injure each other – or one of the farmers. Socially distanced for safety's sake. On a count of three they're let off their ropes and everyone stands back to watch the magic happen as they thunderously gallop up, down and across the paddock. "As wonderful as it looks, it's actually quite stressful when you see all of the Shires running about at full pelt," says Robert. "We really really don't want them to run into each other and cause injury. They're such powerful animals and we love to see them enjoying themselves, but it's always with some trepidation. It's a relief for us when we can get them back into the relative safety of their sand paddock."

In May 2021, the youngest member of the Shire horse posse was brought into the farmyard and harnessed to a new training area set up by equine expert Ruth. In a report made for the website, she explained why she'd decided on a new outdoor location for training: "Sapphire needs to learn that sometimes she will be tied up for longer than about ten minutes, so she mustn't start chewing through her ropes and getting petulant. This is important because when I'm grooming her I need to be safe. When she's taller and I'm grooming the back of her, I don't want her to swing around and send me flying."

Sapphire was obviously nonplussed by the experience and started throwing her head around, but farmer Ruth stood firm, giving Sapphire a telling-off when she became agitated. "She needs to go through this and learn to behave and it's all part of her training – that when she's on a rope she can't play." It was a new experience for the young filly and a hurdle to be crossed in her learning process at Cannon Hall Farm. All of the horses are carefully trained, both for their own wellbeing and for the safety of the farm workers and visitors. Don't worry, they're never punished or put under any stress; each Shire still has its individual personality. However, knowing that they are not going to throw their considerable weight around makes for a better relationship for one and all. They are beautiful animals, but their mighty size means that they have to be handled with great care. When there are members of the public around and potential danger to both animals and humans, careful training is a must.

—

From the largest farm animal to the smallest, May is a special month for the pygmy gang at Cannon Hall Farm as there are still plenty of new arrivals on the cards. Because of the pygmies' tiny size, the farmers are relieved that they do not need to assist in their births too often as the goats are so good at delivering their young unaided, but there are often a few that need a helping hand. And Kate's tiny hands are perfect for the job. "Whenever we have a breech birth, it's essential to get the baby out quickly," says Kate. "Pygmy goats are such little fighters that, when they are halfway out, you may even see their little legs kicking to get free."

In early May, the team was worried that a pygmy goat was holding on to her unborn kids for too long – almost as if she wasn't allowing herself to give birth. There was concern that her kids would get too big inside her. The reason was obvious when Kate helped out the new arrivals – not only were there three bouncing new kids, but, in relation to their mum's diminutive body, they were quite sizeable. Nevertheless, even after a long, stressful birth, in just over an hour, the new family of four looked the picture of good health and would soon be off to join the other newbies in the goats' adventure playground.

A few days later, there was another new arrival, as Shakira the alpaca gave birth to a little boy cria, who was later named Robert. But it was touch-and-go whether he would survive. Robert and David managed to milk a few precious drops of colostrum from Shakira, but the cria was so weak he could barely drink it. The first few hours are crucial for crias as some-times they can just give up on life, and unfortunately the little one's condition worsened and he was rushed to the vet.

Miraculously, after an intense two-day stay, little Robert was allowed to go home. But with his mum Shakira not being able to produce enough milk, and all the Cannon Hall Farm goat's milk already spoken for, Robert and David needed to source some more. Luckily, a local goat farmer came to the rescue as he had more goat's milk than he needed from one of his Anglo-Nubians. Anglo-Nubian goats are known for the high fat content of their milk and his nanny goat Dorsey was the answer to the brothers' prayers. Back at the farm, little Robert soon had all the nutrition he needed to start growing healthily.

The farm's Special Care Unit was filled with fresh straw and put back into action and the cria was lavished with around-the-clock attention. Robert took the 4am feeding shifts, farmer Kate and other team members bottle-fed him during the day, and David covered the evenings. "You can't beat experience," says David. "Helen the alpaca was in a similar situation in 2020 when we had to hand-feed her and we discovered that goat's milk really works for situations like these."

Baby alpacas don't need as much milk as a baby lamb, but the team ensured little Robert had as much nutrition as he needed and was always kept nice and warm. Even though he was born in May, it was still very chilly at night on the farm. The farmers were also keen for him to have some company and the perfect new pal soon came along – Annie, the orphaned pygmy goat. Her mum had unfortunately had such a difficult labour that she was never able to recover.

Reassuringly for everyone, bouncy little pygmy orphan Annie and Robert the cute new cria hit it off from the word go and became very lovely, albeit unlikely, pen pals. (That's the living-together, rather than letter-writing variety.)

"After such a touch-and-go start for the two of them, it's been wonderful being able to see them thriving together," says Robert. "Farmer Kate has been amazing with the amount of attention and care that she has given them – even taking them for little walks around the farmyard. Therefore, we are confident that little cria Robert and pygmy goat Annie will be with us for a long time to come."

———

"In May, we start shearing the gimmer hoggs," David explains. "These are the lambs that will have been born the previous year and haven't yet been sheared. Because they have so much compacted wool on their bodies, they are more difficult to get the clippers through, so they are harder than most to shear. Saying that, because they haven't yet given birth to lambs, they're not yet milking, so there's less chance of causing them any harm. The danger with shearing a milking ewe is it can potentially damage their teats and therefore the supply of milk for their lambs. We shear the gimmer hoggs early in the season and get that job out of the way before we need the time to get to work on all of the others."

Sheep are never sheared until they are a year old, unless they have fly strike (a fly infestation). "Sometimes you get a patch of maggots and you shear off the maggoty fleece, but, if you miss even just a tiny bit, the maggots move along the sheep and infest it again," says David. "You have to keep going back to check that the fleece is completely clean."

Lambswool comes from freshly sheared gimmer hoggs, so in theory the fleeces from the Nicholsons' one-year-olds should be a valuable bi-product, but unfortunately this is no longer the case. At one time, the wool cheque from a flock of sheep would cover the cost of the farm's rent, but not any more. Wool sales have been in decline for decades. "The wool is usually gathered up and sold to the Wool Marketing Board, rather than individual sellers," David explains. "When we get the professionals in to do the bulk of the season's shearing, the money that we make from their fleeces doesn't even cover the cost of shearing the sheep. But it makes the sheep more comfortable in the warmer months, so we do it for the sheep's welfare more than anything else."

"Once you take the heavy wool off a sheep, it has an excellent chance of really enjoying the summer," says Robert. "There's nothing worse than a sheep that's as hot as hell; it just won't thrive. If we can get them sheared by the end of May, we can also get the big problem of fly infestation out of the equation too." Alarmingly, sheep can die within just 24 hours of being infected by maggots.

To create a sheep shearing station, an area is cordoned off in the Roundhouse where the shearers can set up their equipment and have a batch of sheep waiting to be clipped. Some professional shearers have their own trailer fitted with a type of saloon door and in/out system. Then it's just a matter of collecting a sheep, switching on the clippers and away they go.

"We start the shearing at the May bank holidays in the Roundhouse so that the public can watch us at work," says David, who is the chief sheep shearer between the two farming brothers. Robert's not so keen. "I learned how to do it at college and I've done it a few times since, but it really isn't for me," he says. "I let Dave get on with it so that he can show off how good he is!"

Robert's kind like that.

"It's exhausting work," says David. "First you have to catch the ewe, which can be a job in itself as sheep naturally run away from you if you approach them. Once you've caught your sheep, you flip it over and drag it to the right position to start shearing. If it's a big sheep it can be a bit of a wrestle. You can be out of breath before you've even started."

With the ewe on her bottom and her back towards the shearer, the animal is held in position by the shearer with one arm, leaving the other free for shearing. The powerful clippers are connected to a motor, which drives the shears, while a comb underneath the shearing blade glides along the sheep's body and separates the fleece from the skin. "It's not at all stressful for the sheep," says David. "It's actually more stressful for the shearer because you have to lean over them and constantly keep moving, so it's a real test for your back. Although some sheep try and battle to get away, others will just sit on their bum and accept the experience. You even get the odd one licking your arm because they quite enjoy it. It's an intense few minutes while you are shearing each one, but seeing them spring up into the air when they are so much lighter and bounce off back into the barn is an absolute joy."

Some breeds of lamb are easier to shear than others, depending on whether the ewe is a first-time shearer (a gimmer lamb) or an older ewe that's used to being shorn regularly. A gimmer lamb has thick wool all over its tummy, which means it takes longer to move the clipper blades through it, but older ewes grow less wool on their bellies each year. "You get certain breeds like the Dutch Spotted that are easier to shear," says David, "but the thick wooled sheep like the Swiss Valais Black-nose are horrendous to work with – plus they have to be sheared twice a year! My personal favourite is the Soay as that breed doesn't need shearing at all as forty per cent of their coat is hair and it drops off naturally. It's handy for the birds on the farm as they can line their nests with it."

As for timing, that varies too. "A crossbred sheep with no wool on its tummy can be sheared in around three minutes," says David. "Others can take around five minutes each, so I shear an average of forty in an afternoon if all goes well. It takes a lot longer when the wool is difficult to separate from the skin when it's heavy with lanolin. Sometimes you can't even get your clippers in to shear it, and when it's late in the day, you just feel like giving up."

Roger used to shear all of the Cannon Hall Farm sheep and his preferred technique was to go down one side of the sheep, then the other side, then the back of its head and down its back, rather like the sheep was wearing a cape, but David has his own favoured method: "I start at the back of the sheep's head, go on to the front leg, then on to the belly, pushing any wrinkles away with my knee to straighten the belly out, then I'll do the back leg, stand up straighter, push down and do the other leg and up along the back. That way, it comes off as one full fleece. Job done. Dad showed me his way of shearing but I didn't like it so I designed my own method. And, by modifying his method, I've cut a whole minute off the time it takes to shear each one. And that adds up when you have a whole field of them to do!"

"Dave isn't classically trained in shearing," says Robert, "but he has a good technique that works. It's also kind to the sheep. Commercial shearers are paid a certain amount per sheep and they can power on through the flock, whereas when we do it ourselves, we can maybe be a bit more considerate towards the animal and take it a bit more slowly. It can look a bit cruel because you have to wrangle the sheep round to get them in position, but we try and make it as sympathetic a process as possible."

As each sheep is sheared, fleeces are rolled up, sorted into colours, then bagged up and put into huge sacks the size of a skip for the Wool Marketing Board. There, it's weighed, washed and graded for what it might be used for. Coarse wool from fleeces such as the Swiss Valais Blacknose and the Herdwick sheep might be spun into carpet cloth or used for insulation, and although each sackful is huge (think of a bag stuffed with several duvets), it's only worth a few pounds.

"In 2021 we decided to try and sell our fleeces with their provenance attached," says Robert. "We get the best-quality wool from our Jacobs and Kerry Hill sheep, so we bag up each individual fleece and take a picture of the sheep that it has come from, so that the spinner can see exactly which individual has provided the wool they are working with. We get much more money from each fleece this way plus it feels that we are helping to increase the link between the farmer and the craftsman."

The thick, waxy lanolin that coats the wool and acts as a super-charged water repellent is also a useful bi-product from shearing and can be used to make all manner of cosmetics and hair products. It's gloopy and greasy and is sometimes called

Wool Fat, so it's not the most pleasant substance to work with. "You have to wait until the time is right, because if you shear sheep too early, you just can't get the clippers through because of the weight of the lanolin," says David. "When I know I'm going to have to shear sheep that have lots of lanolin – like the gimmer hoggs – I always wear an old pair of jeans that are ready for the bin. I'll wear them every day for the duration of the shearing and by the time I've finished they look like waxed trousers – and they'll be even more ready for the bin!"

Filming *This Week on the Farm* means that David isn't always around for the whole of the shearing season, so professionals like Joe Smith make the prospect of shearing Cannon Hall Farm's 500 sheep a lot easier to contemplate. "Thankfully, Joe shears the majority of our sheep," says David, "but shearers like him are hard to tie down because they are in so much demand. It's backbreaking work and shearers really earn their money as it's probably the most labour-intensive job on the farm. They're a tough breed. But, like so many farm tasks, it's vital that shearing is done well, so we'll always try and get the very best people to do it."

24

When they were younger, the Nicholson brothers loved to watch dad Roger in action when he was shearing, hoping to one day get a chance to have a go themselves – possibly on each other's heads. One day when the brothers' grandad Ted was visiting from Halifax, he jokingly asked Roger for a haircut. "Off Dad went with the clippers and he sheared a massive stripe across his head!" David says. "We were all laughing our heads off, but it was a right mess and it must have really hurt Grandad Ted as well because there's a hell of a pull on those clippers!" No prizes for guessing where the Nicholson brothers got their cheeky streak . . .

Back then, if there was mischief to be made, the three Nicholson brothers would make it. With the weather being warmer, May was a time for being outside as much as possible; riding bikes, playing football, making dens, fishing and thinking up challenges for each other.

"We'd invite friends up from the village and re-enact the seventies TV show *Superstars* on the front lawn of Cannon Hall Museum," says Richard. "We'd regularly get chased off there by the head gardener Mr Hales, who regarded his lawns as sacrosanct. Games of any sort were frowned upon and, at the time, as far as we were concerned he appeared to live a fairly joyless existence based largely around preventing our fun. He and my dad were like chalk and cheese and never got on, but you couldn't fault Mr Hales's dedication to keeping the gardens in tip-top condition."

"They were good boys together, but they took a bit of watching," says Cynthia. "I could always tell when they were up to no good because they'd give me 'that look', or they'd go really quiet at dinner time, then I'd know for sure that they were hiding something."

One May tradition that certainly used to get the thumbs down from the boys was dancing around the maypole. Its origins are sketchy, with some saying it's a Germanic pagan fertility festival possibly as old as the Iron Age or early Medieval cultures. Others claim it's a British festival that began in medieval times to celebrate the beginning of spring. But one thing's for sure, it wasn't a tradition that the Nicholson brothers were ever happy to join in with.

"We loved the annual Mayday festivals," says Cynthia. "The whole village used to come together and there would be dancing around the maypole and the crowning of the May Queen. It was a lovely family day."

David is not quite so keen on the memory. "We had a maypole outside the village hall and I had to do maypole dancing and I hated it," he says. "You had to wear a silly hat and it was really tight around your chops and you had to do this dance where you all held ribbons and one child went one way and another child went the other way and you're meant to end up with the ribbon all twisted into a lovely pattern. I remember Robert and Richard killing themselves laughing at me from the sidelines because it was always the youngest family member who had to do it. They used to shout, 'Look at Dancing Dave!' and I'd be there, prancing around and looking like a right chump."

—

Now life is about making new traditions. All three brothers have children of their own, who were all as mischievous as their cheeky parents when they were younger. Two of whom have May birthdays.

"Tom was a December baby," says Robert. "He had a difficult birth and he had to be incubated for a long time. If he'd been a lamb, you wouldn't have fancied his chances. On the other hand, Katie, our May baby, had a really straightforward arrival. Julie had been to the cash and carry in the morning and I'd been at the wood yard. When I came back at lunchtime, Julie calmly said, 'Right, think you better take me to hospital.' We just got in the car and off we went and Katie was born that afternoon at five o'clock."

But when David's daughter Poppy came along, it was a bit of a drama. "Anita was booked in for a caesarean and the day before the op, we were driving up to the hospital in Anita's car," says David, "when all of a sudden the clutch went and we broke down. I said, 'I'll push and you steer,' but she jumped out of the car and started pushing. Well, there were all these cars going past us, looking at this heavily pregnant lady pushing the car while I was just sat in the driving seat like Lord Muck, and you should have seen the looks I was getting! Anyway, I think that pushing the car may have set things in motion because the following morning, Anita was bumped up the queue for a much earlier caesarean.

"I was back home and didn't know the time of her op had

changed, so I thought I'd get a few hours' work done before I headed to the hospital. But at 6am I got a call from the hospital saying Anita was in labour. I rushed over there and it was very stressful because there were about twenty people in the room, all gowned up and looking concerned. All of a sudden I saw this great big leg appear out of Anita's tummy, then the baby's head got stuck and it felt like a long time until I heard the baby cry. I felt terrible not being able to help because if it had been a lamb I would have been straight in there. Then, once Poppy was born, I felt guilty as I got to hold her for a long time while Anita was being stitched up. I felt bad that she had done all the work and I got to be the first one to hold the baby.

"They were both safely home a few days later on the bank holiday Monday. In hindsight, I suppose I should have taken that day off to welcome the two of them back – but I knew it was going to be a busy day!"

As the saying goes, "like father, like son . . ."

David can always be relied on to come out with a cheeky remark or a bit of banter to lift everyone's spirits. In May 2021, he thought he'd have a crack at training Noddy to be a sheep goat. After all, Robert has Pip the wonder dog, so why not try to emulate his sibling's shepherding prowess?

In a brilliantly silly film for the farm's website, David showed just how easy (or not) it is to train a young pygmy goat. Using a rather natty dog's lead, David encouragingly said, "Sit!" Noddy didn't. Dave tried again: "OK, walk on then." Noddy tootled off, this way and that, with David encouraging him. "That's it! Come by! Away! I really think this is going to be the original sheep goat. Noddy is going to be ten out of ten at rounding sheep up." The fluffy pygmy certainly looked enthusiastic enough, wagging his little tail and enjoying the fun, but when David started saying, "Noddy, lie down!" the pygmy just looked plain confused. David admitted defeat: "Oh 'eck, I've got me work cut out."

Meanwhile, Pip, the farm's official sheepdog, is turning out to be a very sound investment. In autumn 2020, Robert and David travelled to Northumberland, where Robert bought his beautiful new Border collie Pip. It would take four years for her to reach full maturity, so plenty of time to enjoy her as a puppy and teach her the basics. Throughout spring 2021, Robert had Pip at his side, training her to take living on a busy farm in her stride. She needed to get to know the sights, sounds and smells around her so that she wouldn't try and herd up the first animals that she saw – Elvis the llama would probably give her short shrift if she tried that anyway. And she soon started getting to grips with the shepherding commands.

"Pip had some time off from training during the very busy lambing time in 2021 because it's important that the sheep stay nice and settled," says Robert. "But it's vital that I maintain her training, so that she knows how to lie down properly, go left, go right and bring the sheep back towards me. As the summer progresses, I'm hoping to advance her skills, up her pace and make her the sheepdog that we know she can become."

The clever collie is only too happy to please Robert and, although she's still a work in progress, she's continuing to show excellent potential. Some of the rare breeds live outdoors all year round, so there are always sheep to practise rounding up – even the odd billy goat or two. "Soays and Herdwicks will naturally run away when disturbed, which is ideal," says Robert. "The last thing you want is for a sheep to pick a fight with a sheepdog."

With Pip's fine lithe body primed for action, Robert ensures he calmly calls out instructions rather than shouting at her, but sometimes he needs to be quite firm to stop her getting too excited. "Pip is still a novice – she's far from the finished article," says Robert. "But she's got the basics, she knows left and right, and she knows 'Lie down'. She's generally well behaved.

It's just a case of giving her plenty of experience with the sheep and ensuring that she gets the time to learn. Hopefully she will continue being a real asset to the farm, plus I really need a sheepdog now as David's not as fast on his feet as he used to be. Or as obedient!"

—

In the middle of May 2020, Cannon Hall Farm was able to announce some very exciting news – that the farm had been added to the list of the Shire Horse Society's approved visitor centres. "It's absolutely amazing," said farmer Ruth at the time. "It's something that I have been working on for the last year. We'll be one of only eight centres in the UK that have been approved by the Shire Horse Society and part of that responsibility is that we educate the general public about the Shire horse breed – what they did in the past, how they have been challenged now and what the future holds for them. Being an approved visitor centre means that we can work more closely with the Shire Horse Society and we can work together with the seven other approved centres to ensure that this iconic breed doesn't die out. Visitors always ask about the horses as they are often featured on Robert and David's television shows and we can keep bringing their stories to the public via the TV and social media."

Shire horses have become an increasing part of the breeding programme at Cannon Hall Farm, ever since the birth of beautiful foal Will in spring 2020. Working closely with Bill and Sarah Bedford's stud farm near York in North Yorkshire, the Nicholsons have taken several of their stunning Shire horse mares to stud there and it's hoped that there will be several more Shire foals born on the farm in years to come.

It's a magnificent sight, seeing Cannon Hall Farm's collection of farm animals enjoying stretching their legs out in the fields. From contented cows chewing the cud, a dozen breeds of sheep peppering the rolling Yorkshire landscape, beautiful donkeys and Shetland ponies through to alpacas, llamas and the odd goat all out in the open air living their best lives. Spring has sprung and it's smiles all round. As Robert says, "What could be better than walking around the countryside, taking it all in? The grass is a vibrant green, everything's new and fresh. Lambing has all finished, the bad weather has gone. Roll on summer!"

PART SIX
Here Comes Summer!

25

The year-round breeding programme means that every month there are animals giving birth at Cannon Hall Farm. In the spring, it's virtually every single day, so the farm swells massively in numbers. Because it can be so busy, the farmers try to schedule arrival dates around the lambing seasons, ensuring there's always something new for visitors to see. It also means that there will be members of the farming team ready to spring into action should any assistance be needed during any of the animals' labours.

In midsummer, the cattle are given their second routine PD (pregnancy diagnosis) of the year. While they are all out enjoying the sunshine, it seems a shame to disturb them, but it's an essential step, so the first task is rounding them all up. Any farm animal is likely to head in the other direction if you suddenly run towards it, so slowly and steadily, the herd is gathered together by a trio of the farmers, walked to the sand paddock and, one by one, filed into the handling system. The vet then inserts the scanning probe right up the cow's backside (long plastic gloves are essential!) and the results of what the probe sees are transmitted to a portable screen. From looking at any movement and shaded areas visible through the walls of the womb, the vet can see if there's any evidence of a developing calf.

In summer 2021, there was a new member of the Anniversary Herd to welcome, as one of the heifers gave birth to a chunky new white bull. White Shorthorn bulls have a special significance at Cannon Hall Farm as they are very close to Roger's heart. The White Bull restaurant is a celebration of the prize-winning Shorthorn bull Roger's dad Charlie used to own and Shorthorns have always been Roger's favourite. "We didn't have any Shorthorn cows on the farm for many years and I've really enjoyed having them back," says Roger. "They are a dream to see in the fields because of the range of variations in the breed's colour. So much more interesting than just black and white cows. They're such a nice, quiet, well-behaved breed, and since they have been calving so well, it's been wonderful to have them back on the farm."

"Admittedly we had a bad start with the Anniversary Herd," says Robert, "and it took a while for them to settle in, which is unusual for cattle, but, since their arrival in February 2019, we've had around thirty calves and have got into a nice routine with them now. We have built up the herd with some of Jeremy's daughters, so we need a new bull to service them as Jeremy is moving on."

But never fear, Jeremy fans! It was decided that he could have a second career as a stock bull at a neighbouring farm. "Philip Penrose, who does our haymaking for us, was looking for a new breeding bull for his two hundred dairy cattle," says Robert. "Jeremy is going to be kept very busy – if anything, it's a great move for him as it's certainly better than just having eighteen girlfriends. It's with a heavy heart that he is moving on, but he's only going to the next village and will have a great future with one of our pals, who we know will look after him."

Later in the summer, the brothers started making plans to

buy an early Christmas present for Roger at the biggest bull sale in Great Britain. They knew that the bidding was likely to be fierce and that they would need to hold their nerve and grit their teeth to ensure they bought the perfect new pal for Roger's beloved Shorthorn heifers. After all, Jeremy is a mighty act to have to follow.

There was to be another sad goodbye in summer 2021 as Brussel and Sprout, the beloved Swiss Valais Blacknose sheep, also went on to pastures new. "We took them along to Blacknose Beauties, which is a beauty pageant for Swiss Valais Blacknose sheep held in Carlisle every year," says Robert. "We managed to get a first prize for Sprout and a third prize for Brussel, which I was a bit miffed about as Dave was showing Sprout and I was showing Brussel!" These two never let up when it comes to competition! "We'd been thinking that Brussel and Sprout should have a chance at siring sheep of their own, so we found new owners for them at the show and they went on to secure bright futures at adoring homes. It was greatly rewarding – the culmination of a year's effort to save a breed that, at one time, was nearly dying out. Now it's going from strength to strength."

It was difficult to have to let Brussel and Sprout go because they have become such favourites with visitors to Cannon Hall Farm. "Not everything can stay on the farm," says Robert, "and it's not fair to keep them here because they deserve a breeding future. This way, it's the best of both worlds: we've had a wonderful time with them, and we can send them off with our hearts filled with joy, knowing that they are going to continue to have a great life."

Summertime in the countryside means show season. In Yorkshire, the big daddy of them all is the Great Yorkshire Show, which has been held every July since 1838. When the brothers were growing up, they spent a day at the show every year – even being allowed a day off school to go there. "When I was growing up, nearly every village had their own show," says Roger. "It was a big part of summertime. You'd get the catalogue to see what animals would be showing, then decide which of your own farm animals you wanted to put forward to compete. You'd then spend the next few weeks getting the animal into really tip-top condition, training them to walk on a halter, giving them a few baths as needed and scraping their horns with a bit of glass to make them shine."

It was – and still is – quite an undertaking to get an animal show-ready and it now costs considerably more to compete in a top show. "Back then, it was just a few shillings," says Roger, "and most weekends throughout the summer, my father and I would go to a show. I remember once I took a white Shorthorn heifer to compete in a show, while Dad was at another with a roan Shorthorn bull. His bull didn't win anything that day and my heifer won the overall championship!"

Meanwhile, over at the other side of Yorkshire, young Cynthia and her brother Ted would also be off to country festivals, often competing at the annual Halifax Country Show with their prize-winning rabbits. Cynthia has also taken home prizes for her baking, sewing and wine-producing skills.

With plenty of farm produce the whole year round, Cynthia is no stranger to creating an interesting vintage or two, and in the past has created tipples from virtually all of nature's bounty. Rhubarb is a spring favourite and, in the past, she has made barley wine, wheat wine and even a vintage from mangel-wurzels (similar to a turnip). "A friend said I'll bring you some mangolds," says Cynthia, "but I wish he hadn't done. It was awful!" At one time, Roger's sister Shirley was a greengrocer, and when she had produce left over, Cynthia would turn it into wine. "I'd have demijohns of all sorts brewing away in the airing cupboard and there'd be no room for the boys' clothes," says Cynthia. "Some of it was nicer than others and Roger would get quite tiddly on it, but I never touched the stuff!"

"She was too busy drinking gin!" says Roger.

—

When they weren't preparing for a country show, summertime, when Roger and Cynthia were growing up, often meant seaside holidays with the family. "Mum's sister Flo lived about a quarter of a mile from the Blackpool front," says Roger, "so twice a year we'd make the three-hour journey up there, playing car games to while away the time, such as who could spot the most pubs, who'd be first to see the Tower, that kind of thing. Anything to try and distract me from my car sickness!"

Once they arrived at Roger's auntie's, there'd be donkey rides on the beach, trips up the Blackpool Tower, a visit to the zoo and a night at the circus to see world-famous clowns Charlie Cairoli and Paul. "It was such a magical time," says Roger. "I went to the Pleasure Beach so many times and spent so much money there I could have bought it! My favourite thing to do was a game called Monkey on a Stick and I used to win it quite a lot, which wasn't bad going as there were twenty-two of us competing. The trouble was, the prizes were nothing to write home about and we'd be stacked out with rubbish at the end of it – sets of cheap glasses and toys that were worth next to nothing, but I treasured them.

"I also have a vivid memory of going to the paddling pond in Blackpool with my dad when I was really tiny," Roger says. "I'd been given a toy motorboat and I couldn't wait to try it out. I set it off over at the paddling pond, but it got stuck in a jungle of pond weed. I was really upset and thought I'd lost it, but my dad rolled up his trouser legs and waded valiantly into the middle of the pond to bring it back for me. I can still see him now, in his suit, waistcoat and tie and pocket watch, gingerly making his way to the boat. It was really greasy and slippery and my mum and I held our breaths in case he slipped and went under. Everyone was crowding around and laughing, but he didn't see the funny side at all. He wasn't best pleased with me."

Cynthia used to holiday at her grandma's house in Blackhall, County Durham. She and her brother Ted would go up on the train with their bikes so they could cycle from the station to her house. From there, they could walk to the beach. "It's much nicer now than it was then," says Cynthia. "The beach was close to the colliery and they used to tip the waste coal into the sea. People would go down and gather bags of broken coal and sell it on. Ted and I would go to Crimdon instead, which was the next beach along and much nicer. There was a huge ballroom there and an amusement arcade and fairground. Ted and I would take our bikes down there and once a year the whole family would have a reunion on the beach. Because Mum lived a long way from the rest of the family in the north-east, we'd arrange a day and time, and hire a tent that we could all meet up in. No phones back then, so there'd be no cancelling if the weather was awful and it was bucketing down with rain. We'd just all meet up and have a family catch-up. We'd take a picnic and buy cups of hot water to make our own drinks."

Years later, when the boys were all young, Cynthia would take them up in the car for family visits. "The boys all went down to the beach one day to meet their cousins and they decided to build a raft," says Cynthia. "Robert managed to drop the whole thing on his foot and he still has the scar to this day. He survived, though."

26

When the brothers were younger, if they weren't packing their fishing rods for seaside mini breaks, they could always find fun things to do on the farm. Their childhood friend Jayne Bailey lived in the curator's flat at Cannon Hall while she was at primary school and recalls how even back then the boys would all compete against each other – nothing new there then. "Robert was always the most competitive boisterous one," says Jayne. "I don't think he ever forgave me for getting into the football team before him. And a girl getting into the football team back in the seventies was quite a thing! He often managed to beat Richard when they competed, but he couldn't beat me!

"Meanwhile, Richard was always the quiet one. Whenever he disappeared, we'd find him at home in front of the television – he was mad about children's TV, and David being the youngest would always be trailing around after us and whining, 'Can I play?'"

Jayne remembers the masses of rhododendron bushes at Cannon Hall Farm (they're still there to this day) and how she and the boys would make dens there and climb the trees around the Cannon Hall Park estate. "I was a city girl originally, so coming to live on a farm was an absolute revelation. I knew nothing about animals or where food came from, and I didn't believe Robert when he told me how eggs were laid. He made me follow him around the farmyard until we saw one in action, which was quite an eye-opener!"

From playing cricket on the lawn in front of Cannon Hall, to building the float for their annual summer carnival, there was always plenty to do and loads of space to play and disappear in. Jayne also recalls Richard's childhood love of newts and toads and discovering them in a disused toilet near the farm. "Richard always liked creepy crawlies too and we'd go looking for them in the outbuildings, but I never wanted to get too close to David's pet snakes," she says.

Decades later and quite by accident, Jayne spotted the Nicholsons on TV. She then reconnected with them online after more than 40 years. "I was looking for an old school friend via the social media site Friends Reunited and Richard and I started catching up. It was wonderful to hear how the family has transformed what was once an ordinary small farm into the success story it is now. Roger and Cynthia always took me under their wing when I visited and made me feel so welcome. They have worked amazingly hard to make a living from the farm as it was always quite a struggle back then."

"We spent endless days with Jayne," says Robert. "I remember she used to hammer me at conkers and she was a very influential part of our childhood."

The gang all went to Cawthorne Primary School together and would also hang out at the local youth club in the village hall. "The organisers did their best, but they only had five records including 'Kids in America' by Kim Wilde and 'Young Parisians' by Adam and the Ants, which we played in rotation to death," says Robert. "We'd have outings to ice skate at Silver Blades in Sheffield, but mainly it was just hanging out in the village hall."

Meanwhile, David didn't really come out of his shell until

he started college, when he and his mates would stay up all night partying in summer and get up to all sorts of mischief, such as seeing how many of them could get into a VW Polo – nine, it turns out.

"I know I was quite a late starter," David says. "My college years were very different, though! But maybe that needs to be kept for another book!"

—

Back in the present, David might still behave like a big kid sometimes, but when it comes to the main task of summer on the farm, he's one hundred per cent committed.

"With the gimmer hoggs sheared early, we'll tackle the rest of the flock from June first – or whenever the weather tells us," he says. "We get the job done as early in the summer as we can because then it means fewer flies. It also means less chance of a sheep getting stuck on their back and not being able to roll over because they're weighed down by heavy fleece."

As mentioned earlier, David used to shear the sheep single-handedly, but now, with a much bigger flock, he has more help. "We also have to ferry the sheep back to their fields every night because we don't want them away from their lambs for too long. So it's great to have a big team of us working on the task."

The llamas don't need to be sheared as their coats drop away naturally and they shed whatever wool needs to be shed. "Like the discarded lamb wool, birds take the llama wool for their nests," says Robert. "We like to think that our birds have the most varied nesting material of any birds in the UK. It's something we're really proud of."

The alpacas, however, need a good short back and sides every summer. "We shear them once a year," says David, "apart from the little crias, which we leave until the following year. It's important that we shear them because they get very hot and start slavering so it's essential for their wellbeing. But they don't really like it – and they do like to spit, so it's not a great spectator sport!"

In the first series of *This Week on the Farm*, beautiful alpaca Shakira got her very first shearing from Robert. "David told me that I shouldn't shear her head, so I left bobbly bits around her head and her legs," says Robert. "It wasn't a hundred per cent successful, though; she looked like she'd been in a poodle parlour!"

"And he took so long shearing her that by the time he finished, the blades were blunted," says David. Robert has since left the job to the experts.

Another summer maintenance task which is essential for the animals' wellbeing is vaccinating the sheep against potentially miscarrying their lambs. Like any of the other routine health sessions, it's a team effort to vaccinate each and every one of the sheep and it can be a very long, arduous day, whatever the weather. Every single one has to be rounded up and each dose is individually administered, injected into the muscle of either the neck or the leg.

"There's nothing more heartbreaking to a shepherd than nurturing sheep all the way through to lambing time, then, two weeks before they're due, they start to lose them," says Robert. "It's dispiriting – especially if your flock is your livelihood. Thankfully, we've got wonderful scientists who keep coming up with solutions to these problems and we are forever in their debt."

—

Everyone loves a change of scenery and that's the same for the animals at Cannon Hall Farm, with the exception of the pygmy goats and the reindeer, who tend to be badly affected by worms if they go out into the fields. But most of the other four-legged residents love the feel of grass under their feet. The cows will be happily chewing the cud, the sheep will be keeping the grass trimmed in other fields, and the llamas and other animals also enjoy spending time in the sunshine.

"We don't just move animals around for moving's sake," says Robert. "We continually rotate the sheep in the fields because a clean field will not have built up a worm burden. If we didn't do this, the sheep would ingest fly larvae from the grass that hatch inside their intestines and that's what causes all the problems. Once excreted, their poo will further infect the grass and, if sheep continually feed on the same grass, it's a vicious circle. But if the field is rested for several months and cut for hay then it's clean grass again."

However, there are certain times when the farmers don't want the sheep to eat too much rich grass and strive to keep their nutrition on an even keel. "Then, when it's nearly time for mating, we'll give them an extra boost of nutrition. This is known as flushing," Robert explains. "It encourages the sheep to ovulate more and that way they are likely to give birth to twins and triplets as opposed to just single lambs."

Although the Shorthorn cows love to spend summer in the fields too, any cows that the farmers know are due to give birth will be kept back in the farm buildings. "It's impossible to keep a close eye on them when they are out in the field," says David. "If there was to be a problem, it would probably be too late by the time that someone got to them. So we're not holding them back to spoil their fun, it's for their protection."

Like the heavily pregnant Shorthorns, some of the Shetland ponies will also be kept indoors to stop unborn foals getting too big inside their mums. But for the rest of the Shetland gang, they'll get to stretch their legs on the fresh new grass.

And there's one day in the calendar that's extra-special for the Shetland pony posse – the day when Ozzy Horsebourne gets to share a field with Alice, Pony M and the other fillies. First, the fillies are lead to the paddock, calmly accompanied by the farming gang. Meanwhile, feisty little Ozzy – who for the rest of the year lives with the other Shetland bachelors Jon Bon Pony and Pony Hadley – is harnessed up ready to greet the ladies. And it doesn't take long for him to cotton on to the fact that the same event happened last summer – and that he's in for some fun.

"We always expect fireworks," says Robert. "But if the little mares aren't ready to mate, Ozzy Horsebourne may get a badge or two." Alice, the head of the filly pack, will be first in line to tell the young stud whether she is in season or not, by either kicking out at Ozzy or submissively keeping still so that he can mount her. Then the farmers release the other fillies so that Ozzy can continue to romance any of them that might be amenable. Manes fly as the Shetlands race each other up and down the paddock, before settling down for some fresh grass and, if they feel inclined, some summer loving . . .

"We're so grateful to Ozzy Horsebourne for being the missing link in our Shetland fertility programme," says Robert. Jon Bon Pony was the original Shetland stud at Cannon Hall Farm, but, although he sired pony Hadley with Pony M, when it came to the next date to mate, it seemed he'd lost his mojo – hence the addition of Ozzy. "Poor Jon Bon tried his best but he just wasn't fertile enough to do the job," says Robert. "However, I feel duty-bound to give him another chance to see if he can get a tune out of his wonderful old piano one last time and get another one of the Shetland mares pregnant. You never know – where there's a will, there's a way, and where there's life, there's hope. So Jon Bon will still get to spend time with some of the ladies."

The fruits of Ozzy's ongoing labours were plain to see in summer 2021, when two of the fillies gave birth to beautiful young foals. One was called Honey, named after the first horse owned by presenter Jules Hudson, and the second one was given a name to continue the pop star naming tradition – Cheryl Foal.

27

One of the quintessential farming jobs of summer is haymaking: cutting and collecting grass that will be used for animal feed. Straw comes from dried crops and is used for bedding.

First, the long summer grass needs to be cut, and, with so many acres to tackle, it's no use relying on the garden Flymo in the shed. It's time to bring out the big guns. When the Nicholsons first bought Cannon Hall Farm, Roger attached a mower to his tractor to tackle the fields and the baling would be handled by contractor Eric Ellis. But, in the first series of *This Week on the Farm* in 2019, Rob and David had a go at making haylage, enjoying the feel of the massive machinery at work cutting the grass, then returning four days later when the grass was dry enough to bale. "Making good hay is vital," says Robert. "When we open a bale in February, it needs to smell of the sunshine and ideally it will be a mixture of leafy clover and grasses such as Timothy, Rye and Meadow Fescue."

Haylage is cut earlier in the season to other hay and is left in the field for a shorter period of time, then it is baled and wrapped in several layers of plastic. As it has less time to dry out, there will be a certain amount of moisture content which causes it to ferment, but it needs to be sealed up so that no air gets to it to make it mouldy. "Making haylage means that you can bale it more quickly so you are not losing any of its nutritional value," says Roger. "It's like putting it in a tin – its goodness is sealed inside and it comes out the way it goes in."

Silage is made in a similar way, but cut earlier in the year and preserved while it is still green rather than being allowed to dry out. After it has been harvested, the crop is shredded and compacted into a silo (or clamp), where it ferments into a pickle-like feed for farm animals. Unlike fresh-smelling hay, silage has a much stronger "farmyard" aroma. "Silage is slimy horrible stuff," says Richard, "and, although the animals like it, on a farm like ours that's open to the public, it wouldn't go down well." There's no point in creating the authentic farmyard experience if it's unpleasant.

With the weather so changeable, when the farmers are making haylage, it's often a race against time to get the crop baled, wrapped and stacked. A little bit of rain can have a knock-on effect for the whole year if there is nothing to feed the animals, so it's quite literally "making hay while the sun shines". A spinning rake device called a hay tedder is attached to the back of the tractor to collect the hay into long neat rows and then another attachment compresses the cut grass into bales. Bales are then stacked into piles before being wrapped in thick plastic using another ingenious farming gizmo.

Hay is cut later in the year, when the grass is dryer and is ready to be immediately stored in a barn. Traditionally it was a real family affair for the Nicholsons. "In the past, it would be the three of us lads on a trailer loading the hay bales," says Richard, "with my mum driving the tractor and Dad chucking the hay bales up to us." Once they had a trailer load, it would be all back to the farmyard to get it unloaded and up into the barn, using an elevator belt to get it into the store from the ground level. "The elevator belt was completely open, there were no guards or anything like that," says Richard. "It's amazing that we survived uninjured really. Dad would be directing operations and chucking the bales in all different directions, and me, Rob and Dave would be positioned at various points so all was evenly stacked. Dad was very handy with the hay fork and made light work of chucking the bales to us. It was flipping hard work though – a real workout – and it would take a day at least to do each field." All this was post-mechanisation, when previously, the grass would have had to be scythed by hand. (Think *Poldark* . . . No, not like that.)

It sounds like a simple enough process, but there is a science to making hay. "If you bale it when it's not really ready you can see the steam coming off it in the barn and you know you are in a bit of trouble," says Robert. "You know that it's going to go mouldy and that the feed value will greatly reduce. Also, you can't make hay until the pollen is on it," he explains, "and it needs to be at the right point for all the grass varieties in the field to shed their seeds so the field can continue to be plentiful. If the timing is right, we can get a few nice cuts of hay throughout the summer. There's a point where if you leave it too long, it will have all gone to seed. It's still useful but it won't be its absolute best. So it's about assessing the optimal point where the pollen is set, and while some of the varieties have started to seed, it's still predominantly leafy, and that's when it should be cut and turned into hay."

Before the brothers were old enough to help out with the haymaking, Roger and Cynthia would be aided and abetted by their friends Rosemary and Nigel. "I remember wearing my bikini top while haymaking," says Rosemary. "It was always late summer by then so I could top up my tan at the same time! Roger and Nigel would throw the bales on to the trailer and Cynthia and I would stack them ready for the farmyard. We'd do thousands of bales between us and it always felt like it took a few weeks. It was great fun though and we'd always celebrate the end of it with a party!"

—

Not all farming tasks are quite such good fun, like the annual dentistry for the alpaca herd. And when it comes to tricky customers, they don't come with more of a challenge than headstrong Zander, aka Robert's nemesis. When he gets his annual once-over, he needs all the strength of two farmers to hold him in place, while a vet gets to work. For his 2021 checkup, it was Yorkshire vet David Melleney in the driving seat: "I just needed to file down Zander's teeth a little. There are no nerves in alpaca teeth, so it doesn't hurt him, but the electronic rasp does sound a bit alarming."

With the rasp plugged in and the potentially most difficult patient out of the way, it was plain sailing with the rest of the alpaca pack. Audrey would usually be next in line as she has a history of her bottom teeth getting too long, but in July she was busy giving birth to a beautiful new baby cria, who was later named Dolly. Fans will recall that Helen, Audrey's first cria, had to be hand-reared because Audrey rejected her, but Audrey instantly bonded with her second newborn in July 2021. "Little Dolly is doing amazingly well," says Robert. "She is growing fast and she's really beautiful. It's really life-affirming that her mum Audrey has the chance to bring up such a lovely cria after the disappointment she had last year with Helen. Seeing her with her own baby, living her best life is absolutely wonderful. It's the highlight of 2021 for me!"

On to the next summer task: giving the tups their foot vaccines. In a film for the farm website, David explained why it's such a vital part of their ongoing health: "It keeps the sheep sound on their feet all summer, so it's something we like to do. Back in the day, we'd be turning over the sheep and trimming and spraying their feet around five times a year, so the new vaccine has reduced that necessity considerably. The trouble is, if you inject yourself with the vaccine, you can lose a finger because the solution is so strong!"

With 11 tups to vaccinate, the farm's latest handling system is put to good use. It's a fairly simple procedure as David works on each one of the sheep in turn, injecting just below one of their ears – while keeping his own fingers totally clear of the needle. "There is one negative," he says. "The injection does cause a little lump as a reaction that we have to keep an eye on to make sure it doesn't turn into an infection, but the positives far outweigh any negatives."

"Lambing is probably the busiest time on the farm," says Farmer Dale, "but over the summer months there's prolonged work outdoors making sure that all the sheep are in tip-top health, so the four hundred and thirty ewes and all the lambs will get their foot vacs too. The tups will be breeding again in autumn time and any problems in their feet could affect their ability to serve the ewes."

28

With summer well and truly here, and with plenty of happy and healthy new lambs, plus calves, crias, piglets, baby donkeys, Shetland foals and all manner of other newborns, the farmers have more time for celebrations. And, in summer 2021, there was a lovely reason for the family to get together and raise their glasses, when David's daughter Poppy married her fiancé, Henry.

"Despite the pandemic and the restrictions on the numbers allowed at weddings, Poppy wanted to get married as planned that June," says David. "When Poppy and Henry got engaged, they decided the wedding would be five hundred days later and so we had a small wedding of thirty people and a big shindig a month later. My best man from my wedding in Jamaica came along. I haven't seen him for twenty-six years, but I'll never forget that date – partly because it's on the same day as my birthday, the twenty-second of April! My top tip – get married on your birthday and you'll never forget an anniversary!"

Another highlight of summer 2021 was the christening of Rob's granddaughter Nelly, which had been delayed somewhat because of the Covid-19 pandemic. "It was the most lovely christening I have ever been to," says proud Great-Uncle

David. "Because it couldn't happen when Nelly was a tiny baby, she couldn't wear Katie's christening gown that had been made from Julie's wedding dress, but at least Katie got a picture of her in it when she was a baby. On the day itself, Nelly was running around, shouting her head off and throwing her teddy everywhere. It was really funny."

Between the wedding, christening, country show commitments and the inaugural Five on the Farm Festival, all taking place in summer 2021, there was hardly any time to draw breath. But it's all part of the farming cycle at Cannon Hall Farm.

"There's always the next thing, then the next thing," says David, "but it's been bred into us. January is all about maintenance on the farm, fixing stuff that needs it and getting ready for February lambing. Then there's lambing itself in February and getting ready for Easter lambing. Through March there'll be kidding too and spring cultivation to organise and, before we know it, we're into the shearing season, which is the next busiest time after lambing. When lambing is over, we can usually get our sleep back on track, but in 2021, little Rob the alpaca came along and had to be hand-fed throughout the night, so the end of spring pretty much merged straight into summer. That's the thing about farming: so much is dependent on the weather. Then there are animals that come along that need to be cared for and, before you know it, another month has passed.

"When all the bluebells are out and there are lush colours everywhere, you can enjoy the spring in all its glory and it just can't be beaten. You can relax a bit more when spring is over, but there's always something to plan and always so much going on – sheep tupping, the Shire horses to get in foal, the pumpkin festival to organise, and then Christmas. There's always something, but it's always so rewarding. It's the best job in the world."

Learn the Lingo

A few useful farming phrases

animal husbandry – The science of breeding and caring for
farm animals

bagging up – When the teats of a female animal fill with milk
prior to giving birth

breech birth – When the legs are born first rather than the
head

corn – The generic term farmers use for cereal crops,
including wheat, barley and oats

cria – A baby alpaca or llama

crutching – Trimming the fleece around a ewe's bottom to
make mating easier

drilling – The farming term for planting crops

farrowing – Another name for pigging

flushing – Increasing the ewe's nutrition to encourage
ovulation

gilt – A female pig is called a gilt until its second litter; from
then on it is known as a sow

gimmer hogg – A lamb that has been weaned but not yet
sheared, usually around a year old

heifer – A cow that hasn't yet had a calf, or has only borne one

in bulling – Term used to describe when a heifer is in season
leveret – A baby hare
pullet – A young female chicken
raddle – A harness that transfers dye from a ram to a sheep
during mating to indicate which ram has served each sheep
scour – Another name for diarrhoea
shearling – A sheep that has been sheared once
side door – A term vets use to indicate a caesarean
springing to calf – The sign that birthing is imminent as the
cow's udders show signs of filling with milk, plus other
signs that she is in the later stages of pregnancy
stockmanship – The art and science of properly handling
cattle and other farm animals (stock)
tup – A farming term for a ram
tupping – A farming term for mating sheep
yearling – A horse that's a year old

Meet the Team

There's a whole army of people who work with the Nicholsons at Cannon Hall Farm, all helping to make it Yorkshire's number-one farm attraction. Here are just a few of them.

Wade Pass, farmer

When Roger had to step back for a well-earned break in spring 2021, Wade was in the right place at the right time to take over his role in charge of the pigs at Cannon Hall Farm. At just 25 years old, it was no mean feat to take on such a big task, but considering Wade had already been farming for eight years, he was well up for the challenge: "When I left school at sixteen, I had no interest in going to college – and, to be honest, I probably would never have stuck it out," he says. "I had always been interested in farming and then I heard about dark training, which seemed the ideal way to learn on the job."

Dark training is a way of getting qualifications while you gain practical experience, and Wade managed to secure employment at J & E Dickinson, a large mixed farm in Yorkshire. It not only had dairy and beef cattle, but also reared pigs and grew arable crops, so Wade could experience practical farming skills across the board. In order to become fully trained, his mentor would visit him at the farm once a month to assess his assignments and set next tasks to be completed. In just a few years, Wade was able to secure qualifications in Livestock Production, Livestock Management and Business Management. By the age of 21, he was a fully fledged farmer, with experience working across all the departments of a mixed farm.

Taking two years out of farming to drive heavy building machinery in the Peak District, Wade enjoyed the work but wanted to get back into working with animals in the countryside. While he was back working at Dickinson's, his mentor from his training years told him about a new opening at the Nicholsons' farm. Wade's mentor is John Hopkinson, who is the Health, Safety and Compliance Officer at Cannon Hall Farm, and when Robert and David said they had a gap for a talented farm worker, John immediately suggested Wade.

"When I came to meet them to talk about taking over from Roger, I was really impressed with the set-up of the farrowing facilities," says Wade. "The fact they are so well designed and ventilated, and there are plenty of escape routes for the piglets." (Escape routes, by the way, are the barriers in place for piglets to get out of the way of the heavy sows so that they don't get squashed, not so that they can escape into the big wide world . . .)

Wade is no stranger to hard work, and when he's not in the farrowing barns, he will be working alongside other team members with the cows, sheep and other animals. And with his experience of handling major agricultural machinery, he's always on hand to teach David a thing or two about his JCB driving. "I love winding him up," says Wade, "but I know who's boss . . ."

Dick Dickinson, builder

"I've been at Cannon Hall Farm for nineteen years and it's the fastest nineteen years of my life," says "Spotted Dick" Dickinson. Back in 2002, Dick and his father, Stuart, were employed to do a little building job at Tower Cottage, the Nicholsons' original home at the farm. Since then, the building team have been employed full-time, working on houses, shops, farm buildings and everything in-between. "When I left school, I went straight to work for my father, who was a hard, hard man," says Dick. "He didn't tolerate anything less than the very best, but it's stood me in good stead because Robert runs a tight ship and he's a perfectionist too. It's all about perfectly straight lines with him."

The first major project that Dick worked on was a two-year build renovating Mill Farm in 2004 into a habitable home. "It was an absolute ruin, so we had to pull the whole lot down, but, as it was a listed building, we had to work within certain parameters." His team also knocked the old farm down at Cannon Hall, put it through a crusher and built it up again.

Dick and his team then converted another local farm into several ultra-modern houses for Robert and David, complete with underground cinema rooms. Back at Cannon Hall Farm, he pulled down barns and turned them into state-of-the-art facilities such as the Hungry Llama restaurant. Then there was the amazing Roundhouse project. "For that we had to remove thousands and thousands of tonnes of earth to get to a level plateau, but the building fitted like a glove once we got the foundations in," he says. "Then the roof went on like a massive umbrella."

More huge building projects followed for Dick, always with his fantastic workmates Colesy and Kieran by his side. The pig unit, the rare breeds barn, the farrowing houses . . . Virtually every building that you see on the farm has been built by Dick and the team. "We've added to the original café twice, built the Lucky Pup twice and we've finished the small mammals house and the squirrel enclosure," says Dick. "There's always plenty of work to do and we're given a list of tasks every day by Robert and David. They don't let the grass grow under our feet."

Dick has become such a fixture of the farm from being on Robert's Facebook Live broadcasts that he is now recognised when he's out and about locally. "Robert started calling me Spotted Dick and now people want to have their photograph taken with me. It's just weird," says Dick. "I'm not one for the limelight though; I'll just slink away."

Alex Booth, farmer

As dedicated Cannon Hall Farm fans will know, the team love to ask their Facebook followers to vote on a name for a new animal, and when their feisty male alpaca arrived, it was a choice between Dave (after David) or Zander, after young farm hand Alex. "The supporters went for Zander and I'll always have that one up on Dave," says Alex, who since then hasn't had any other animals named after him. There's still time, of course.

Alex came to work on the farm in 2013 as a parking attendant when he was 17. Back then he was at college studying English, Psychology and Business Studies, and he worked on the farm at the weekends. But, as for so many other staff members at Cannon Hall Farm, what started out as a little part-time job grew and grew. In 2021, Alex had clocked up eight years as a fully fledged, full-time member of staff. "I started to do little jobs on the farming side," he says, "but I actually thought I would get fired on my first day because when I was mucking out the llamas, one of them got out." He obviously got a second chance as he is now one of the longest-serving members of the farming team.

Spring is Alex's favourite time of year on the farm because he can help out with the lambing, but it's Petal the pygmy goat that tops his list of farm favourites: "She was born in 2018, and farmer Ruth and I had to really fight to save her. She was absolutely tiny – even for a pygmy – and we really didn't think she was going to make it. We were so delighted when we were able to save her, and now she's probably the friendliest animal on the farm."

You might recognise Alex from the film reports he makes for the Cannon Hall Farm website and if you've visited the farm, you may have seen him presenting one of the special animal talks. "We change the subjects of our talks from season to season," says Alex. "In spring, we do lambing talks, then, as we move into summer, we'll talk about the Shire horses and how to groom them. Later in summer, it's all about the meerkats and mongooses."

Alex is also to be heard commentating on the sheep and ferret racing: "When we first opened up again after both of the lockdown periods, we weren't able to do any animal racing because of social distancing, so when we returned to the talks and the commentary, we were all a bit rusty at first. We had to re-learn how to do it, but it's a great laugh and gets everyone involved. We even have a pretend bookies so we can come up with odds for who's going to win."

When he's not perfecting his comedy commentary, like several of the other farmers, Alex likes to be behind the wheel of the JCB, driving into the barns and topping up the hay feeders. He's also become adept at mucking out with the JCB: "I prefer it to mucking out by hand, that's for sure!"

Rob Hampshire, carpenter

You might recognise Rob Hampshire, Robert's son-in-law, from the *On the Farm* television shows as he and his workmate Ash can turn their hand to anything that needs building on the farm – from the Hungry Llama to the Reptile House, the gift shop and the new Mammal House, plus the brilliant playgrounds for the pygmy goats and the stylish hen house. The Shire horses and reindeer also live in the green oak paddocks built by Rob and Ash. And, during Covid-19, Rob used part of his furlough time to build a beautiful greenhouse for grandmother-in-law Cynthia: "For as long as I've been here, Cynthia has wanted a greenhouse, so I just got a load of wood and glass and built it for her as a bit of a surprise."

Rob always wanted to be a joiner, ever since he helped his dad rebuild a rusty old speed boat he bought while on holiday in Wales. "Dad tinkered with the engine and I took all the paint off," says Rob. It looked so good that his dad then bought two or three more to do up, plus a couple of motorbikes and several caravans. "We weren't a wealthy family so if we wanted anything, we would buy a fixer-upper and work on it. I remember going into the local DIY shop when I was about seven years old and loving the whole environment, the smell of the wood and all the power tools. From the age of about ten, I started to pester them to let me work there, and, at thirteen, they gave me a Saturday job. I absolutely loved it, and worked every day of my school holidays there (and sometimes the odd school day!), and by the time I went to study carpentry at Barnsley College, I could use pretty much every machine."

With the nuts and bolts of a great career ahead of him, Rob first worked for a local company called Nicholson and Roberts (not knowing, of course, that his future father-in-law would have a spookily similar name) and had his own successful carpentry company for years before heading off to Australia for five months.

"Katie and I had just got back from travelling and one of the first things Robert said to me was, 'I've got a couple of weeks' worth of work for you, can you start on Wednesday?' Seven years on and I'm still here."

Rob's wife, Katie, is the head of HR at Cannon Hall Farm and their daughter Nelly is the apple of everyone's eye. "It's a great family," says Rob. "It's what I'd describe as the dream family because, although we all work on the farm, everyone is trusted to get on with their job. It's what these guys all do for each other that makes it work so well."

Georgie Kaye, farmer

Georgie's very first day at Cannon Hall Farm in June 2021 coincided with a day of filming of *Summer on the Farm*, so she was well and truly thrown in the deep end. "It was a Saturday and it was manic," she says. "People were flying around everywhere and it was crazy busy, but that's the best way to learn, isn't it? I knew farmer Ruth already so she showed me around the farm and I went straight into mucking out. It was a case of 'this is what we do, crack on!'"

Georgie was the perfect new recruit for the farm as she has bags of experience looking after both farm and zoo animals. As she's from a farming background, she has always had animals around her, and she also studied Animal Conservation Science at the University of Cumbria. While she was there, she saved up enough money to spend two weeks volunteering at the Bornean Sun Bear Conservation Centre. "Their main purpose is to rehabilitate sun bears that have been kept illegally as pets, and some are actually kept as ingredients for restaurants. It was an incredible experience seeing the rehabilitated bears being safely reintroduced to the rainforest."

Following her ambition to get a salaried job at a zoo, Georgie did unpaid work experience at a number of great wildlife centres and zoos around the UK, including the Welsh Mountain Zoo, Chester Zoo, Tropical World in Leeds and the Yorkshire Wildlife Park, working with everything from giraffes, polar bears and tigers, to penguins, pine martens and otters. She then got a job as a keeper at Wigfield Farm, which is part of Barnsley College, where farmer Ruth also used to work. At that time they didn't work together. "Ruth hired me for my job at the college and we stayed in touch, which is how I ended up at Cannon Hall Farm."

After three and a half years, Georgie got her job at the Nicholsons' farm, and all those years of training with both farm and zoo animals are paying off: "I love it here, because I get to use all my farming skills and my skills on the wildlife side. And, because of my zoo experience, I can help Kate with the exotic animals in the Reptile House, and assist Ruth with the zoo licensing. We're in the process of finalising which species will be housed in the new Small Mammals House, so it's been really exciting getting involved with that too. It's nice to work on a project right from the start, and that was a big draw for me coming here."

Georgie particularly likes working with the mongoose collection, who are nearly as new to Cannon Hall Farm as she is. All in all, it's a big thumbs up so far: "I'd seen bits of the *On the Farm* programmes and when I was asked to join the team, I watched a lot more so that I knew what was going on. There's not many other places you can get to watch on television before you go along for your first day! I have to say that everyone is so welcoming and exactly the same as you see them on TV. They're brilliant to work for."

Charlotte Ellis, farmer

If Charlotte's surname sounds familiar, it's because her grandad Eric Ellis worked for the Nicholson family for many years as their agricultural contractor, only retiring from the farm in 2021. Now, Charlotte is following in Eric's footsteps, having grown up on the family farm just two miles away from Cannon Hall Farm.

Charlotte's family have a substantial herd of dairy cows and have a flock of pedigree Ryeland sheep, several of which have won prestigious awards. Therefore, Charlotte is well used to every aspect of lambing and sheep farming. In 2021, she took two of Cannon Hall Farm's Dutch Spotted sheep to compete in the Royal Cheshire County Show. "It was the first time that Cannon Hall Farm had shown sheep and it was really good fun," she says. "Especially as they were both pet lambs at the farm, meaning we'd reared them by hand so they were really friendly. Everest the lamb is my favourite animal on the farm."

Charlotte also fell in love with the pygmy goats so much that she adopted two butterscotch-coloured cuties for herself in summer 2021: "I decided to call them Spud and Smoky and, when I first got them home, I put them in our small pen and I nipped to our straw shed to get some little bales to stop them escaping. I was only gone for thirty seconds and when I came back, Mum was holding on to Spud, who'd already got out. They're certainly a lot more boisterous than sheep . . ."

Charlotte studied at the prestigious Askham Bryan farming college near York before taking on a very unusual part-time job as a dairy cow hairdresser. "It wasn't a job I'd heard of either! It was really good fun though – trimming up cows around their legs and bellies, udders and around their lovely faces. Not to make them show ready, but to help keep them clean before they wintered indoors."

From February 2019, Charlotte has worked full-time at Cannon Hall Farm: "I love it because I get to do so many things," she says. "I do quite a lot on the public side and do a lot of the school tours, which I really enjoy; it's rewarding, and challenging too, but in a good way. Having such a variety of animals that I'd never worked with before is wonderful – especially animals like alpacas and goats. It's kind of crazy to walk in and see them all running around, but that's why it's such good fun."

Seasonal Cooking

by Tim Bilton

Chef Tim Bilton has worked with Cannon Hall Farm since 2014 and hosts the "Cooking with Tim" slot on the series On the Farm.

As a chef, the changing of the seasons is such a delight for me because it means new foods to enjoy at their freshest and most plentiful. Spring is one of my favourite times of the year. After the bleak winter, the air starts to become warmer, birdsong lifts the mornings, and life on the ground, in the trees and in the bushes starts to burst into action.

In the shops, you'll still find lots of fantastic winter fruit and veg, such as apples, beetroot, potatoes and, of course, sprouts. With sprouts, I like to slice them thinly and quickly stir-fry them with a handful of cubed pancetta. I make this for my boys every Christmas and they can't get enough of it. Or try a lovely winter pud of Bramley apple and foraged blackberry crumble. Add rolled oats to your crumble and a few tablespoons of ground almonds for an even tastier topping.

Old-fashioned favourites like hearty winter stews and game pies still taste good to this day because they can be made from local, fresh ingredients. If you shop locally, it's better for the environment because there are fewer land miles involved in getting your food from the farm to your plate. Before people had access to food from all over the world, and prior to everyone having frozen food, it was always about what you could eat that was fresh – nowadays, we're just better at getting ordinary food to taste extraordinary!

Just after the peak of the root vegetable season, and before the bountiful months of spring greens, gardeners call March the hungry gap because it's a time when the ground is not heaving with produce. Nevertheless, you can still cook seasonally. Have a look at what's freshest and the best value in the greengrocer because that's what's likely to be most plentiful – and therefore in season. Get a load of onions, carrots, squash and celeriac, chop it up and sweat it down in some butter and maybe a sprig of rosemary that you leave whole. Then add a litre of stock and simmer for 20 minutes. Remove the rosemary, give it a quick blend, maybe add a drop of cream, season, and you've got a tasty, comforting winter soup.

When the garden starts to wake up, there are some great British gems coming into their seasonal best. Spring greens, Savoy cabbage and purple sprouting broccoli can all be cooked in my special recipe that I think will revolutionise the way you cook vegetables. Due to the speed of the process, the vegetables will retain their colour, flavour and much of their nutritional content. Simply heat your pan to the highest temperature, then cook sliced vegetables (200g is enough for four people) in 50ml of boiling water with a good knob of butter, a tsp of sea salt and a grind of pepper. Put the lid on and cook for between two to five minutes, depending whether you are cooking leafy greens or heavier veg. The cooking liquid with the butter and seasoning form a light emulsion, which has a delicate sheen and a rich taste. Delicious.

As we turn to April and May, the produce arriving at the kitchen at Cannon Hall Farm is full of fresh flavours. There are sweet young shoots, new potatoes and spring lamb to enjoy, and it's a month for lighter dishes that suit the brighter days. It's nice to see things coming to life and you see people with a bit more of a spring in their step. I really love the delicate colours and young, fresh flavours of seasonal food at this time of year. Try vibrant baby carrots cooked with orange, thyme, caramelised garlic and a little splash of honey. My favourite is new season asparagus – Yorkshire is the best, by the way; ask any chef! It's fantastic simply pan-fried, or try roasting it with lemon and garlic and then crumble over some feta cheese. To trim asparagus, simply clip off the bottom woody part and peel a couple of centimetres off the stalk if it's particularly stubby.

Springtime risotto makes the most of spring peas and beans, and I always think that a good risotto that's well made is like getting a big cuddle from someone that you love. I know many people don't share my love for broad beans, but if you pop them out of their wrinkled white jackets, they are really fresh and delicious. If you're cooking them as a side veg or eating them raw, they'll literally need 30 seconds in a pan of boiling water, then immediately plunge them into ice-cold water to keep their vibrant green. Add to salads or pile them on to sour-dough bread topped with cream cheese or ricotta. They're also great to chuck in with creamed leeks.

Foraging is very fashionable now and if you can access some wild garlic and have it with homemade pasta – it's easier to make than you think – the flavour is wondrous. The best things in life really can be free! Spring lamb, however, can be expensive but it's well worth it for an Easter treat, and there's nothing bet-ter than roast lamb and new potatoes with fresh mint sauce – just chop up a handful and add a splash of wine vinegar. Perfect.

—

For an instant home-menu makeover, follow these quick tips:

1. Update your herbs and spices at home to maximise the flavours in your cooking. Better still, try growing your own.
2. Always try and use the freshest, best-quality ingredients you can afford.
3. Slow cooking is fantastic for cheaper cuts of meat and is a lot easier to do well than trying to, for instance, fry a steak or a piece of fish perfectly.
4. Keep spring asparagus upright in a glass of water like a bunch of flowers. Set the bottoms in a few inches of water and cover the buds with a little plastic bag. This will keep it fresh for a good five days.
5. Embrace seasonal food at home by varying what you eat as you go through the year. Experiment with recipes you see on TV and online; don't be afraid to try something new. You never know, you might like it!

Meet the Yorkshire Vets

Peter Wright, Julian Norton, Shona Searson, Matt Smith and David Melleney are the key vets featured in the On the Farm *series, plus everyone's other favourite Channel 5 show,* The Yorkshire Vet.

Peter Wright

Long before *The Yorkshire Vet* was first aired on Channel 5 in 2015, young vet Peter Wright started his career at the same practice where Alf Wight, aka James Herriot, and Donald Sinclair, aka Siegfried Farnon, also worked. *All Creatures Great and Small* was a TV phenomenon when it was first shown in 1978 and over 40 years later, Peter is continuing the tradition of bringing tales of veterinary life to TV.

As *The Yorkshire Vet* is made by the same television production company as *On the Farm*, you'll see many familiar faces popping up in both programmes. After all, they both focus on animal life in Yorkshire . . .

Peter remembers appearing in the very first episode of *Springtime on the Farm*, on a bitterly cold night in April 2018. His wife, Lin, was keeping watch from the sidelines, doling out blankets to those who needed them, and she lent fellow guest Gloria Hunniford her gloves. "I remember actor Kelvin Fletcher being there and he was this big strapping muscle-bound fella," says Peter. "Lin saw how cold he was so she sat him on her knee and wrapped him in a rug."

It was to be the start of a great friendship between all those involved in the two programmes, and, since then, Peter has been back to Cannon Hall Farm on numerous occasions to help with lambing, calving and other springtime tasks. He per-formed surgery on Prince the reindeer in 2020 and went along with Roger, Robert and David to buy Fern the Highland heifer at auction in 2019. "When Roger was bidding, his finger would flick up to indicate he was interested," says Peter. "But as the price started to rise, Roger's finger became less and less mobile! Meanwhile, Robert and David were on the other side of the ring, encouraging him to bid higher."

Peter and Roger have become good pals from working on TV together – once famously eschewing chef hats during a cookery demonstration in favour of their favourite flat caps. Collectively, they have over 100 years of farming experience between them and neither has any plans to retire.

"The whole veterinary world has changed and there are so many things that you can do now that you couldn't do then, such as giving animals MRI scans and treating them for can-cer," Peter says. "I have day books from when Alf/James Herriot used to work, which list his jobs, like 'visit cow – mastitis', and at the end of the consultation there'd be a spread of food for you and a nice Windsor chair by the fire. Now you're lucky to get an outside tap to wash down your waterproofs."

Nevertheless, Peter continues to enjoy working with all creatures great and small from a mixed practice in North York-shire. "Treatments have come on leaps and bounds and even though I'm sixty-five, I have plenty of work in me yet. Besides,

I'd miss the characters that you meet in farming, the day-to-day banter and the challenges of the work."

Julian Norton

When it comes to memorable stories involving the Nicholsons' farm and vet Julian Norton, they don't get more colourful than the story about Gary the donkey's penis. The unsightly growth obviously wasn't much fun for Gary, but it certainly cemented the working relationship and friendship between Julian and the brothers. It's just one of several occasions when the three have found themselves in front of the TV cameras, in both the *On the Farm* programmes and *The Yorkshire Vet*. "My first impression of Rob and Dave was that they were just two normal blokes," says Julian. "I'm from just up the road in Castleford, which has a similar mining heritage to where they live in Barnsley, and we hit it off immediately. I always feel quite connected with them because they ended up on telly by accident in much the same way that I did, so I've always felt that we are kindred spirits in that respect. We have a similar outlook on life."

In spring 2018, Julian had been invited to talk about being a vet on the first series of *Springtime on the Farm*, when his professional skills were suddenly called for in order to rescue a lamb with a prolapsed womb. "After that, whatever issues they had with the animals, Rob and Dave would send me a photo and ring me up for advice, then I'd head down the M1 to see them."

Until 2020, Julian worked at the same veterinary practice as Peter Wright. Being in the heart of farming country in North Yorkshire, springtime was as busy a time for them as it was for the farmers. "Over the years, I recall some absolutely exhausting springtimes," says Julian. "Working for four weeks on the trot with no days off including weekends is very hard. But all the stress and fatigue is mitigated by the fact that there is new life coming into the world, the days are getting brighter, and there's new growth in the fields. It's always a difficult time for anyone connected with livestock, but it's a very happy time too."

The pressures of springtime are on the back burner now as Julian mostly works with domestic animals, which is handy as the family has a Jack Russell called Emmy and a rabbit called Boris. "He's floppy and gormless and rather stupid. So the name Boris seemed quite apt."

The Gary the donkey story made it on to Channel 4's *Gogglebox*, where a panel of viewers react to what they see on-screen. It was such a hit that now Julian has become a regular face on the show. "I'll be watching the telly with the family and suddenly I'll pop up on *Gogglebox*, which is a bit weird. But one of the best things about the TV journey – and I'm sure that Rob and Dave will agree – is being able to show what we do on a daily basis with TV viewers. Farmers and vets get a real sense of satisfaction seeing a good outcome, and if you can share that with two million people watching at home, it's the icing on the cake."

Shona Searson

Shona Searson is well and truly following in her mum's footsteps as she too worked for Donaldson's Veterinary Group, based in West Yorkshire. Back then, Shona's mother was one of only three female graduates out of a class of 60. Now, 80 per cent of newly qualified vets are women, and Shona is delighted that she's able to reflect how the profession has changed. "In our job, there's always someone watching you, whether it's a farmer, a pet owner or a colleague, so I'm used to explaining what I am doing with an animal. Therefore, having a camera pointed towards me when I am working isn't all that different."

Before Shona even graduated from Liverpool University, she was offered a job at Donaldson's and is now one of the directors of the Farm and Equine business. "From the age of about thirteen, every holiday I used to get my dad to dump me at a sheep and beef farm in the Lake District," says Shona. "And I learned to ride before my teens, so all my skills have amalgamated into what I do now."

Although Shona does some small animal work (the poorly hamster variety), her first love is working with farm animals. You may have seen her on *Springtime on the Farm* in 2021 when she helped a very large sow with a prolapsed womb. "Believe it or not, that wasn't the worst pig prolapse I've ever worked on," she says. "I had to repair a rectal prolapse once, which was completely disgusting. What made it worse was I couldn't deal with the problem with my full-length gloves on, so I had to put my bare arm up there. It was horrendous."

There are certainly more glamorous jobs, but Shona wouldn't swap hers for all the tea in Yorkshire. "The farm animal side of vetting is either in you or it isn't. It's about the people as much as it's about their animals – sometimes their whole livelihood can revolve around it and you want to help them as much as you want to help the animal."

And Shona is always delighted to work with the farmers at Cannon Hall Farm, especially as she used to visit the farm as a little girl. "We'd come two or three times a year when it was a petting zoo, so it was really exciting when they became clients of Donaldson's. The Nicholsons are so progressive and they have really brought the farm on. It's changed vastly from when I was younger. The staff that work there are absolutely brilliant and you can tell how much they care about the animals they look after. They really believe in what they do."

Matt Smith

Springtime on the Farm viewers first saw Matt Smith at work in 2019 when he had to perform a caesarean on one of Roger's Shorthorn cows. Unfortunately, it was one of the Anniversary Herd that had been struck by a mystery virus and its calf was stillborn. Thankfully, it was a much happier result when Matt was able to deliver young Roger the beautiful little white bull. "I hadn't been working that day and I was at Cannon Hall Farm watching the calving from the balcony of the Roundhouse," says Matt. "Then suddenly people started crowding around a cow and I knew that something was wrong." Luckily, he was able to step forward and save the day.

Even though Matt was new to viewers back then, in previous years he's carried out various procedures at the Nicholsons' farm. Several have taken place in the Reptile House, where in spring 2021 he performed the most intricate surgery of his career. Tilly the tortoise wasn't laying any eggs and hormone injections hadn't been helping, so Matt needed to operate, sawing a square out of the underside of Tilly's shell to make a window to work through. "All of Tilly's eggs were stuck like Velcro inside her and it was really nasty," says Matt. "I removed them, flushed out the infection, stitched everything back up, then replaced the square of shell like a big piece of jigsaw. She was able to recover back to tip-top health."

After Matt graduated from Glasgow University in 2014, he originally planned to work in Canada for a few years, but he was immediately offered a job at Donaldson's Vets in Huddersfield, where he has worked ever since. "Like a Yorkshire homing pigeon, I came back to the county I'd grown up in. It was a job offer that I couldn't refuse as it was the perfect fit for me."

Matt's brother is a dairy farmer in Cumbria and, from the

age of around ten, young Matt used to help with calving every spring. "That, combined with watching Steve Irwin on the Discovery Channel, is where my love of cows and reptiles comes from."

With a history of exotic pets, including rescue snakes, five tortoises and a bearded dragon called Dennis (who is apparently suffering from depression), Matt really is the vet to handle both the biggest and the smallest of patients, and he loves working alongside Peter Wright. "The stories that he comes out with are fantastic," says Matt. "Many of the veterinary tasks are the same as they have always been. And, although the technology has improved so much, ultimately you can still find yourself in the middle of a field in the driving rain in the middle of January with your arm up a cow's rectum. It's still mostly just about getting a smaller thing out of a bigger thing."

David Melleney

The Nicholsons are indebted to all of the vets that have helped them over the years, but one vet is particularly close to their hearts. David Melleney first met David and the rest of the family when he came to do work experience at Cannon Hall Farm as part of his school GCSEs. "The first task of the day is always mucking out the pigs and, as a younger lad, you feel you have to prove you can stand your ground with a huge sow. But they're so strong and there's no way that you can hold them. Trust me, I've tried many times!"

As his ambition had always been to be a vet, at the end of his week's work experience, he was keen to do more. Farmer David was happy to oblige, saying, "You can have a job, but can you start tomorrow morning?" And that was the start of around five years of working at Cannon Hall Farm on Saturdays, some

Sundays and all through his school and sixth form holidays. "Some of my first memories there were during springtime," says David. "I was always amazed when Dave would just come into the barn, jump over the hurdles, wash his hands and off he went. I remember thinking, 'Right, I need to learn how to do that.'"

With lots of farming experience under his belt and excellent exam results, David went to study Veterinary Medicine at Cambridge University, swapping pig muck for academia. "I didn't really know what to expect, being a Yorkshire lad, and it's fair to say that not everyone at Cambridge had a similar backstory."

After six years of study and no fewer than 28 weeks of clinical placements, David could finally call himself a vet. And it wasn't long until he was back at Cannon Hall Farm. "I was excited about going back there to see everyone, but I was also really nervous and I was keen to make a good impression," he says. But no one could have prepared him for what was going to happen.

"I drove into the farmyard and there was this camera team. Rob and Dave were filming *Springtime on the Farm* and they wanted me to check over one of the Anniversary Herd that was calving. One of the filming team said, 'We hope the calf is alive,' and I nervously said, 'Erm, yeah, me too.' There was a lot at stake as there had been problems with the herd, but thankfully I could help deliver the calf successfully."

Since that day, David has been back to Cannon Hall Farm several times and appeared many times on the *On the Farm* programmes, before the show's producers asked David and co to join the team on *The Yorkshire Vet*. "I owe the Nicholsons so much," says David, "from letting me do work experience with them to the rest of my career so far. They deserve their success because they are such a wonderful family and they have worked so hard. It's been a real privilege to be part of the ride."

The Countryside Code
with Rob and Dave

If you fancy a giggle, put the words "Country Code" into the search bar on the Cannon Hall Farm website (cannonhallfarm. co.uk) and you're in for a treat. Robert and a very large, slightly familiar-looking sheep have some tips about how you can make visits to the countryside safer both for yourself and other countryside lovers and animals that you may encounter.

The Countryside Code was first published in 1951, when there was an influx of visitors from urban areas discovering the joys of the national parks. With the introduction of public access to open country and registered common land in 1949, visitors were allowed to discover the joys of walking through fields and farmland that had previously been no-go areas. With their box-fresh wellies, many people were keen to see what life in the countryside was all about, not realising that you need to keep your wits about you, even on a gentle stroll through the fields. The code also advised farmers not to "regard the holidaymaker from town as his enemy".

Thankfully, the modern version of the Countryside Code is a bit less formal, and now it sounds less like the headmaster telling us all what to do. And here, with Robert and David's help, we can all stick to the rules and make the most of the fantastic countryside.

"Even if you don't have the best walking shoes in the world, put your common-sense head on before you head out," says Robert. "For example, if you see a sheep on its back, don't be afraid to roll it over so it can get back on to its feet. That's the kind of thing that you can do, which is really helpful to farmers."

Rob and Dave's Countryside Tips

Respect everyone

- Be considerate to those living in, working in and enjoying the countryside
- Leave gates and property as you find them
- Do not block access to gateways or driveways when parking
- Be nice, say hello, share the space
- Follow local signs and keep to marked paths unless wider access is available

"Imagine if we were to walk into the place where you work," says David, "totally ignore you, wander around like we own the place and then block the doorway so you couldn't get out? It sounds mad, doesn't it? But there are people who really don't care about the damage that one open gate can cause. And, although not everyone wants to chat to absolutely everyone they meet, it's not hard to at least acknowledge someone if you cross paths. And please don't block gates; no one likes to be barricaded in."

Protect the environment

- Take your litter home – leave no trace of your visit
- Do not light fires and only have BBQs where signs say you can
- Always keep dogs under control and in sight
- Dog poo – bag it and bin it – any public waste bin will do
- Care for nature – do not cause damage or disturbance

"Litter doesn't just look awful," says Robert, "it can be very dangerous to wildlife. A dog that's not on a lead can stress, injure and potentially kill another animal, and if you trample across a field, you can damage crops. Think of it as your own lovely garden and treat it accordingly."

Enjoy the outdoors

- Check your route and local conditions
- Plan your adventure – know what to expect and what you can do
- Enjoy your visit, have fun, make a memory

"We've all done it," says David, "set out in the morning when the sun's shining, then the rain's started pelting down, your phone's out of juice and someone gets a blister. We all love a bit of spontaneity, but as the saying goes, 'failure to prepare is preparing to fail'. Taking time to make sure you have a fully charged phone, a mini first aid kit, a map, some water and lots of layers can make the difference between a lovely day out and a day you'll never forget for all the wrong reasons. Like the Scouts say, 'Be prepared'!"

Acknowledgements

The Nicholson family would like to thank our editor, Claire Collins, and all the team at Ebury Publishing for inviting us to share our springtime story. Thanks again to Amanda Stocks at Exclusive Press & Publicity for helping make the project happen and for guiding us along the way.

Once again, our cousin Steve Wilson was a great help in supplying genealogy information, and it was great to be able to include anecdotes from our childhood friend Jayne Bailey.

Thank you to all of the Yorkshire vets who continue to be a great help to us both on the farm and with our various television programmes.

A huge thank you to all of our incredible staff at Cannon Hall Farm whose loyalty, dedication, hard work and passion help us so much every day. We'd like to thank you individually, but we are bound to miss someone out, so a great big collective thank you to everyone.

Thanks to all of our loyal supporters and Facebook followers. The last couple of years have been a very testing time for us all and it's been wonderful sharing our story with you and getting so much positive feedback for all that we do here on the farm.

We would like to thank Nicole Carmichael for her friendship, help and advice in the writing of this book.

Thank you to Paul Stead and all the team at Daisybeck Studios and to presenters Helen Skelton, Jules Hudson, Adam Henson and JB Gill, who have become great friends as well as filming colleagues.

Roger would like to thank butchers John Holmes and Alan Askwith for their friendship, help and advice in setting up our farm shop. He'd also like to mention his great friend and work colleague, the late Nigel Elliot, who is greatly missed by everyone at the farm. He was completely dedicated to the animals under his care. They loved him as much as he loved them.

Richard would like to thank his partner, Clare, and son, Marshall, for their support and encouragement. Also his friends Matt Harris, David Pittaway, Richard Sykes and Rob Newton for their unstinting friendship and companionship over many years.

Robert would like to thank his wife, Julie, for 34 wonderful years of marriage and his son, Tom, and daughter, Katie, for their love and unwavering support.

David would like to express his thanks to his wife, Anita, for putting up with him for all these years and accepting the long hours that farming demands. Thanks are also due to his daughter, Poppy, for her love and support with all things technical. A big thank you to all his friends, both old and new, for their friendship over the years.